Central America: A Natural and Cultural History

Central America

*A Natural and
Cultural History*

Edited by Anthony G. Coates

Foreword by Olga F. Linares

USAID A Publication of the Paseo Pantera Project

Yale University Press
New Haven and London

Paseo Pantera is a Regional Natural Resources Management project conducted by the Wildlife Conservation Society and the Caribbean Conservation Corporation in collaboration with Tropical Research and Development, Inc., with partial funding provided by the United States Agency for International Development. This publication was made possible through support provided by the Office of Guatemala-Central American Programs, Bureau for Latin America and the Caribbean, U.S. Agency for International Development, under the terms of Cooperative Agreement No. 596-0150-A-00-0587-00. The opinions expressed herein are those of the authors and do not necessarily reflect the views of the U.S. Agency for International Development.

The Wildlife Conservation Society, since its founding in 1895 as the New York Zoological Society, has been saving wildlife and inspiring people to care about our natural heritage. Today, WCS conducts the world's leading international conservation program devoted to saving endangered species and ecosystems and manages the nation's largest system of urban wildlife parks, including the world-famous Bronx Zoo. It has helped to establish more than 100 wildlife sanctuaries around the world. WCS manages over 270 field conservation projects in 50 countries, among which are a set of projects in Central America working to protect habitat and eventually create an unbroken greenway along the length of the isthmus—Paseo Pantera.

Caribbean Conservation Corporation is a nonprofit organization founded in 1959 to support the work of the pioneering naturalist and zoologist Archie Carr. CCC is dedicated to the preservation and conservation of sea turtles and other coastal and marine wildlife through research, education, advocacy, and protection of natural areas.

Published with assistance from the foundation established in memory of Philip Hamilton McMillan of the Class of 1894, Yale College.

Designed by James J. Johnson and set in Melior Roman and Syntax types by The Marathon Group, Inc.

Printed in the United States of America by Edwards Brothers, Inc., Ann Arbor, Michigan.

Library of Congress Cataloging-in-Publication Data

Central America : a natural and cultural history / edited by Anthony G. Coates ; foreword by Olga F. Linares.
     p.   cm.
"A publication of the Paseo Pantera Project."
Includes bibliographical references and index.
ISBN 0-300-06829-8 (cloth : alk. paper)

1. Natural history—Central America.   2. Central America—History.   I. Coates, Anthony G. (Anthony George)   II. Paseo Pantera Project.
QH108.A1C45   1997
508.728—dc21   97-11105

A catalogue record for this book is available from the British Library.

The paper in this book meets the guidelines for permanence and durability of the Committee on Production Guidelines for Book Longevity of the Council on Library Resources.

10 9 8 7 6 5 4 3 2 1

Contents

Color plates follow page 78

Foreword

OLGA F. LINARES

Here is a book that recounts history—in the tradition of the *très longue durée*—in the grandest and most illuminating way. *Central America: A Natural and Cultural History* is the account of a biological corridor that took more than 60 million years to develop and is taking less than a century to be destroyed.

The editor-geologist Anthony Coates is deputy director of the Smithsonian Tropical Research Institute (STRI), located in Panama. Coates explains in exceptionally lucid style the workings of plate tectonics. Formed by the collision and separation of five crustal plates riding on a hot mantle, the Central American isthmus had a violent and complex past. When it reached final closure 3 million years ago, the isthmus created a barrier between two oceans and a bridge between two continents. Organisms in the Caribbean and the Pacific increasingly diverged, creating what Jeremy Jackson, a marine paleontologist and senior STRI scientist, and Luis D'Croz, professor of marine biology at the University of Panama and STRI scientist, call "two different ecological realms." The marine faunas on both sides of the isthmus became progressively isolated from each other under fluctuating climatic and oceanographic conditions. In their highly informative yet profoundly disturbing essay, Jackson and D'Croz trace the intricate process of bioconstruction carried out by coral reefs, sea grasses, and mangrove vegetation in both oceans—a process now being reversed as human settlers continue to degrade and destroy coastal environments. The nature writer David Rains Wallace sketches the biogeographical scenario—the vast coral reefs, long coastlines, mangrove swamps, foothills, valleys, and volatile volcanoes, all closely juxtaposed—where multiple dramas unfolded.

Once the isthmus was firmly in place, terrestrial animals from North and South America began their amazing migrations. Echoing his predecessors, the paleontologist David Webb of the University of Florida at Gainesville refers to these movements as the "the great American faunal interchange." Dozens of exotic, previously segregated animal genera—ground sloths, giant anteaters, toxodonts, the gigantic *Titanis* bird—all moved northward and became extinct eventually. Great carnivores and herbivores, many species of deer, and numerous small mice moved southward to speciate and radiate successfully into every conceivable niche. The majestic *Panthera onca*, formerly confined to the southwestern United States, now ranged freely well into northern Argentina. These animals moved through the patchwork of forests, light woodlands, and scrub that dominated the vegetation of the isthmus during the glacial cycles. According to Paul Colinvaux, a STRI ecologist and paleobotanist, the pollen and phytoliths (small silica cells) extracted from the mud of deep lake cores tell of drier and cooler times. Montane and lowland forest tree species struggled to dominate the route taken by the great migrating herbivores.

During all this time, the fate of creatures inhabiting Central America was guided by the free play of evolutionary forces. For millions of years, nature took its normal course. Processes leading to divergences and convergences, extinctions and population explosions gave shape to living forms. Actions and interactions constantly took place along time-space biological and physical boundaries.

Human beings arrived late upon the American scene, at the tail end of the last Ice Age about 12,000 years ago—but they promptly altered the rules of the game. STRI's archaeologist, Richard Cooke, vividly recounts how paleoindian hunters contributed to the disappearance of the large terrestrial mammals. He describes how the descendants of these peoples transformed the vegetation, first by bringing native plants under domestication, then after 2000 B.C. by extensively altering the landscape through their nomadic slash-and-burn agricultural practices. The peoples of Central America who settled south of the Maya may not have built temples and pyramids, but they did produce sculpture, pottery, and gold ornaments of enduring beauty. The Spanish conquest, however, brought an end to all this. Disease, death, and desolation almost obliterated native populations. The Amerindian societies that survived did so by taking shelter in the vast, still-forested territories along the Caribbean coast.

During the 300 years of Spanish colonial domination outlined so competently by Stanley Heckadon-Moreno, STRI's sociologist and human ecologist, the Central American region became unified by a common language, a single faith, and a unitary political system consisting of two *audiencias,* one in Guatemala, the other in Panama. Biotic resources, however, became more diversified, as new cultivars and domestic animals were introduced from the Old World. With the passing of time, the Central American provinces grew apart from Spain until, in the first decades of the nineteenth century, they gained their independence. Their economies, however, grew ever more dependent upon export crops, foreign labor, and overseas investment. The demographic explosion ushered in during the twentieth century resulted in widespread colonization of forested frontiers. These lands were rapidly transformed into vast cattle ranches. Land degradation, increasing poverty, decreasing food security, and political violence predictably followed.

Within this bleak Central American scenario, forty-five distinct indigenous groups numbering 3.9–5.6 million people and constituting anywhere from 13 to 19 percent of the total population presently struggle to protect the reduced natural resources that remain under their control. The geographer Peter Herlihy of the University of Kansas is quite correct when he argues that only those economically sustainable management strategies that recognize indigenous rights over land and resources have some chance of succeeding. Ecologically sound conservation practices alone offer hope for recreating the vital links that once held between Central American peoples, their habitus, and their habitats.

It is in this spirit of bioreconstruction that the Paseo Pantera project was courageously conceived in 1990. As the Panamanian biogeographer and historian Jorge Illueca, deputy director of the United Nations Environment Programme in Kenya, carefully explains, the mandate of the Paseo Pantera project is to safeguard biodiversity. Its goal is to restore to the Central American bridge its previous functions as passageway by creating protected ecological corridors that facilitate the movement of wildlife along its entire length. For many millions of impoverished rural peoples, this initiative promises economic recovery through jobs and services in sustainable agriculture and fishing, ecotourism, reforestation, and agroforestry.

The authors of this splendid volume are passionately committed to making Paseo Pantera a reality. To this end, they bring impressive knowledge, skills, and vision to bear upon a work of reconstruction that has no match in breadth or intensity. In producing a book with a profound and indelible message they forward a dream of social justice and ecological well-being: a peaceful and prosperous world built on a common understanding of the wise use of shared resources.

Preface

ANTHONY COATES and ARCHIE CARR III

This book has arisen from the compelling yet challenging vision of an international conservation project known as Paseo Pantera. Paseo Pantera in most Hispanoamerican countries would be interpreted as "Path of the Jaguar," referring to *Panthera onca,* the large tropical American cat. In Spanish, *pantera* is used for the dark phase of the jaguar and *tigre* for the more common lightly colored, spotted form. The Paseo Pantera project, however, was named by Chuck Carr not for the jaguar but for the cougar *Felis concolor,* known locally as the Florida panther. The cougar ranges through every continental state of North, Central, and South America and thus symbolizes the need for continuity of natural environments throughout Central America if this region's stupendous biodiversity is to be preserved. The Wildlife Conservation Society (WCS) and the Caribbean Conservation Corporation (CCC) took the leadership role in promoting the concept of the Paseo Pantera project as a regional network of wildlife corridors in Central America. The project assisted the nations of Central America in developing a central unifying theme for their collective conservation programs. This book evolved from and was funded by the Paseo Pantera project, which was funded by the U.S. Agency for International Development (AID). As editor, Coates would like to make special acknowledgment to Archie (Chuck) F. Carr III. In addition to his vision and leadership of the Paseo Pantera project, it was his idea to create a book that would tell the unique, compelling story of Central America. Drawing on his vast knowledge of the region, he outlined the components of the book and brought together an original group of authors. He made extremely helpful comments on the first drafts of each chapter and worked closely with the editor through each stage of production. This book would not have been possible without him.

The Paseo Pantera project has had as its purpose the fostering of support for the rehabilitation of the isthmus as a biological corridor between the two continents of the hemisphere. Late in 1994, the governments of Central America committed themselves by treaty to establishing this corridor. The government of the United States, the United Nations Environment Programme, the United Nations Development Programme, the World Bank, the European Community, and other agencies from around the world have agreed to help Central Americans meet this commitment to global environmental security. The Paseo Pantera corridor would reach from the Darién Gap of Panama to the Maya Forest of southern Mexico, Belize, and Guatemala. Since the United Nations Conference on Environment and Development in Rio de Janeiro in 1992, no country or group of countries has made a more imaginative or more hopeful pledge to biodiversity conservation than the seven States of Central America. With this book, we show why that commitment is worth the attention of the world.

To achieve the purposes of Paseo Pantera, to confront the challenges of man and land in the Central American isthmus in an effective and comprehensive way, a reconciliation must be found between three intersecting and increasingly destructive ingredients of Central American life. First, the rising awareness of citizens, governments, and conservationists that Central America is one of the richest biological regions of the planet and that it is currently being destroyed at a terrifying rate. (A recent estimate calculated that the tropical forests of Central America are being cut down at the rate of 88 acres per hour.) The principal success of such groups to date is to have created a series of protected national parks and biosphere reserves in an attempt to sequester some minimum amount of these natural reserves for future generations.

Second, the remaining natural areas of Central America correspond extensively with the traditional homelands of numerous indigenous peoples who have sustainably made a living there for centuries. In many cases the rules of conservation implemented in the parks and biospheres do not have regard for the ways of life of the indigenous peoples, creating the first destructive intersection. Yet stresses on these peoples, modernization of their lives, and growing population pressure have meant that they often no longer live sustainably within their regions, creating another source of conflict.

Third, the burgeoning populations of mestizo peasant farmers, suffering from stunning poverty, lack of education, and lack of agricultural knowledge, are migrating inexorably into the reserves and territories of the indigenous peoples, destroying both the natural resources and the local cultures. This migration is driven by exponential population growth and a mode of agriculture that cannot be sustained because it destroys the very soil it seeks to cultivate. There is therefore an insatiable need for more land.

The intersection of these three elements of Central American culture is leading tragically toward a potentially violent and ultimately catastrophic conclusion for all peoples concerned. Paseo Pantera is a powerful vision to salvage the extraordinary natural beauty and diverse biological riches of Central America by creating a network of corridors within and between countries to connect protected areas and to integrate the natural ecosystems, even across the extraordinary distances that animals such as the cougar need to travel in their natural habitat. Paseo Pantera proposes that these preserved resources form the base of a spectacular enterprise of ecological tourism complemented by sustainable agriculture that uses new techniques and alternative economies for exploiting natural resources. It would be a unified conservation and sustainable development movement for all seven countries of Central America. Surely the sweep and vision, frustrations and successes of such a project are worthy of a carefully crafted book.

Yet as we discussed the very different elements, political, social, historical, and biological, that the concept of Paseo Pantera evokes, we were drawn more and more to realize that Central America is a singular and exceptional region. We also realized that to conserve nature one must first know how the natural system works. That means research. To know what is possible politically and socially involves understanding how the society has evolved from the past. And to use the experience of others we needed to know how similar or different is our own case. We were thus drawn to an even broader sweep of the history of Central America, natural and cultural, one that would provide an appropriate background and setting for the compelling ideas of Paseo Pantera.

And what a splendid experience it has been. Once embarked on such an eclectic project, we quickly became aware that it could not be achieved by one author. We therefore made a virtue out of necessity and assembled the best people we could find for each of the chapters. Each author brought his own critical building block to help construct what we believe is an accurate, comprehensive account for the lay reader of Central America's natural and cultural history.

The editor readily admits that he was not ready for the depth and variety of stories that were originally told by each contributor, far exceeding what could be managed within one volume. This led to very hard decisions about what could be included in each of the chapters and how the chapters should link together. The editor has also tried hard to ensure that this complex review of new and old scholarship comes to the reader at an even rhythm and a constant pitch. He is very much indebted to the contributors for their tolerance of his intrusion into their style and content because it was inevitable that such a diverse array of scholars would vary in the level and extent of detail in their narratives. The book has a vast scope, and no doubt it was risky to tackle such a sweep of history. But Central America seemed poorly described at this popular level; and it does have an extraordinary history, one that is clearly different from the powerful cultures to the north and south.

Furthermore, there are recurring themes in the Central American story that seem to unify the natural and the cultural history in this apparently disparate collection of sagas. The profound differences between the Pacific and Caribbean coasts reverberate from the geological beginnings of violent volcanic versus quiescent plate margins, through strikingly different ocean realms and their contrasting marine life. Culturally, the contrast continues between the older Catholic, mestizo, cattle, and coffee culture of the Pacific and central mountainous regions and the indigenous and criollo (the latter often anglophone and Protestant), forest-dwelling, banana, and reef-fishing cultures of the Caribbean slope. Today, the drier Pacific side is almost completely deforested and overpopulated; the wet Caribbean is still rich, though reduced in forests and wildlife, and it is coveted by those excess populations. Violence has characterized the volcanic geology, the north-south clashes of great carnivores and herbivores at the joining of the land bridge, the conquistadors, and independence. We are today only just recovering from the savage conflicts of the 1980s, the Lost Decade.

We hope, then, that our treatment of the broad spectrum of themes we have tackled crystallizes in the mind of the reader the unique place and role in history of Central America and, while not minimizing the formidable challenges to its peoples, transmits an appreciation of its amazing diversity and exquisite beauty.

Acknowledgments

I had a great deal of help in editing *Central America*. Archie Carr, III, discussed the overall direction and style of the book at each stage of production. He was particularly astute in detecting passages dealing with more technical subjects that were not, as he put it, "user friendly" for the general reader. I am grateful to all of the contributors, not only for accepting my editing of their texts, but for agreeing to review at least one and sometimes two or three drafts of the chapters that preceded or followed their own.

I also thank Archie Carr, III, John Christy, Helena Fortunato, Marcos Guerra, Héctor M. Guzmán, Kathleen Jepson, Nancy Knowlton, Haris Lessios, Ross Robertson, and Zuleika Pinzón for locating, providing, or identifying illustrations and John Pandolfi for kindly creating figure 2–7. The following persons greatly improved one or more chapters through their careful and helpful reviews: Penelope Barnes, Mario Boza, Olga Linares, Ross Robertson, Ira Rubinoff, and Fernando Santos-Granero; to them, sincere thanks. I am also very appreciative of the contributions of the reviewers commissioned by Yale University Press, particularly a careful and thoughtful review by Sally Horn.

Lidia Mann and Gloria Zelaya have greatly helped in tracking communication with the authors and other persons involved in assembling the manuscript. Only they know how much I owe them. Jean Thomson Black, science editor at Yale University Press, saved me with shrewd advice throughout, and the patience and guidance of Lawrence Kenney in the final technical stages of production are gratefully acknowledged.

There are three people whom I simply cannot thank enough for the work they have contributed to *Central America*. Alejandro Caballero is in charge of the Smithsonian Tropical Research Institute's Digital Imaging Laboratory. Often working weekends and evenings, he transformed all of the illustrations into computerized digital images and gave invaluable advice on graphical composition and quality. He was unfailingly cheerful, enthusiastic, and tolerant of the never-ending requests for additions and modifications. Xenia Guerra, my research assistant, also volunteered many hours outside her regular work schedule and was invaluable in her attention to detail and the diversity of tasks she accomplished. Without her, this book could not have been completed. Janet Coates, my wife, not only tolerated many lost evenings and weekends but played a major role in editing my chapter and in preparing the rest of the manuscript. Alejandro, Xenia, Janet—thank you.

Central America: A Natural and Cultural History

CHAPTER ONE

The Forging of Central America

ANTHONY G. COATES

Nowhere in the world does a relatively small sliver of land manifest more dramatically the primal workings of the earth than in Central America. In its geological structure and the variety of its surface expression Central America is one of the most complicated regions on earth. The formation of the Central American isthmus also strongly affected processes operating over a wide area of the surface of the earth. Ocean circulation, climate, and the distribution of plants and animals on land and in the sea were profoundly changed. How and when these changes were triggered by the rise of the Central American isthmus is an area of very active modern research, and the future promises to bring to light new discoveries. For these reasons many scientists believe that the closure of the Central American isthmus was the most important natural event to affect the surface of the earth in the past 60 million years.

Not surprisingly, there is controversy concerning the timing and direction of the complex plate movements that led to the creation of the Central American land bridge. The geological story told here represents the most widely accepted scenario for the geological origins of Central America, but the reader is referred to the references at the end of the book for a discussion of more controversial issues.

The complexity of the geology of Central America is manifest in the variety of its surface terrains, the study of which is known as geomorphology. It is reflected in the asymmetry of geological structure across the isthmus from the Pacific to the Caribbean and in the strong contrasts between the north and south (fig. 1–1). The Pacific side is ge-

1

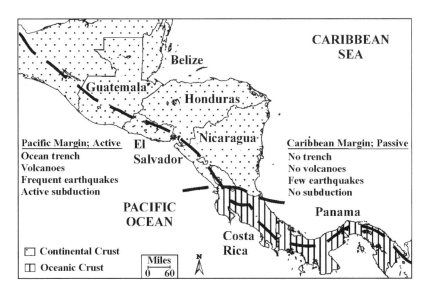

1-1. Map showing the distribution of continental crust (northern Central America) and oceanic crust (southern Central America) and the geological contrasts between the Pacific and Caribbean margins of the isthmus.

ologically active and usually the site of the major earthquakes. The ocean crust on this side is buckled downward near the coast into a trench 2000 meters deep; the volcanic chain tends to hug this coast, and there are vast quantities of sediment eroded from the chain and dumped by submarine avalanches into the trench. By contrast, the Caribbean side is mostly stable, with few volcanoes or earthquakes, no trench, and a more continuous and steady transfer of lesser amounts of sediment to a gently sloping marine shelf.

In northern Central America the land is old; hundreds of millions of years of mountain building and the long, slow process of erosion have sculpted the distinctive limestone karst terrains of El Petén and the granites and deformed metamorphic rocks (preexisting rocks transformed by heat and pressure) of the Crystalline Highlands of southern Guatemala, Honduras, and northern Nicaragua. Volcanic activity has come late to these regions and is superimposed on a broad and ancient rock tapestry. But to the south, in southern Nicaragua, Costa Rica, and western Panama, volcanoes dominate and the landscape is new, built in geologically recent times by seismic and volcanic processes. In eastern Panama, along the narrow Darién bridge to South America, where the isthmus rose completely above sea level only 3 million years ago, the volcanic chain ends and the geology takes on the flavor of the continent to the south.

Geology's Unifying Theory: Plate Tectonics

Why is Central America so complicated geologically? What kinds of processes are at work? Geologists now believe that the inexorable movement and fracturing of the surface crust of the earth, as it rides on the hot, flowing mantle below, create the earth's continuously changing surface features. The process, called plate tectonics, suggested a model for the functioning of the earth that revolutionized geological science in the late sixties.

Plate tectonics is the first earth model to provide a coherent explanation for the location of such diverse features as earthquakes, volcanoes, mountain ranges, and deep linear trenches in the oceans. It also helps locate rich deposits of precious metals as well as major reserves of oil and gas. Plate tectonics was broadly accepted as a theory of the earth only in the 1970s, when geologists finally understood the mechanism by which it worked. But years before, some geologists persistently drew attention to the fact that many biological and geological features did not make sense if the continents had always been in their present positions. Identical species of tiny reptiles called *Mesosaurus* that lived in freshwater swamps 260 million years ago are now found as fossils in Brazil and Africa (fig. 1–2). How did they cross the Atlantic? Rocks that are 250 million years old and of glacial origin are now found in India, South Africa, South America, Antarctica, and Australia, as are fossil leaves of a distinctive genus of tree, *Glossopteris*. Scientists know of no mechanism that could form ice only in those locations nor do they understand how a genus of tree could be a native of five continents and live in so many different climatic conditions.

Paleomagnetism

In 1966 two amazing discoveries about the rock record of the earth's past magnetic field were made. The first was that there is a magnetic component imprinted on certain kinds of rocks when they are formed that allows their original location (latitude) to be determined. Minute crystals of magnetite act as miniature compasses that align themselves, while the rock is still molten or the sediments still dropping through the water, to the lines of force of the earth's magnetic field. At the poles, these lines of force are vertical to the surface of the earth; at the equator they are horizontal. For each latitude in between they are at a unique angle (fig. 1–3). Thus, by measuring the angle of the magnetite crystals and allowing for any subsequent tilting of the strata, the original latitude of the rock when it was formed can be determined. These studies

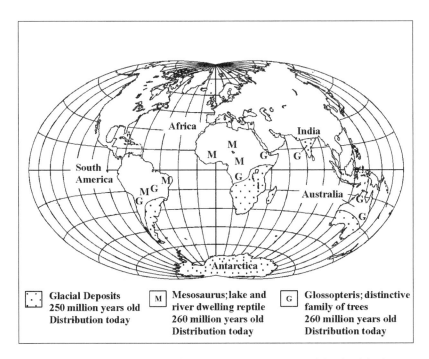

1–2. Map indicating the disjunct and inexplicable distribution of the fossil freshwater lizard *Mesosaurus*, the fossil tree *Glossopteris*, and the glacial deposits of 250–260 million years ago when plotted on a modern global map. After "The Theory of Plate Tectonics," Copyright 1994, Tasa Graphic Arts, Inc.

have shown that most rocks are now far removed from where they were when formed, and in general the older they are, the farther they have moved. Furthermore, measuring of these paleomagnetic angles for rocks of the same age in different continents and then extrapolating where the poles would have been for each continent show that each continent has moved along a different path, that is, each continent produced a different pole position for the same time (fig. 1–4). When continents are moved back so that they have the same pole position for each time interval, for example, 260–250 million years ago, they form a large supercontinent called *Pangea*. The southern contiguous land area within *Pangea* is called *Gondwana*. Here, the rocks with *Mesosaurus* in them occur close together, all the glacial deposits form a single ice sheet covering the pole, and *Glossopteris* occupies a zone at a similar latitude (fig. 1–5).

The second extraordinary discovery about paleomagnetism was that the north and south magnetic poles appear to have reversed themselves erratically during the course of time. At certain times, the paleo-

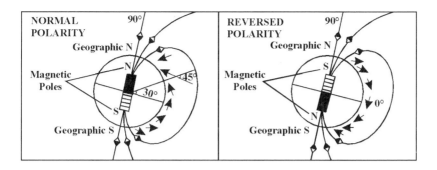

1–3. Diagram of the earth's magnet field during normal and reversed polarities. Directions and polarity of the magnetic lines of force are indicated by arrows. For example, the angle of the magnetic crystals in rocks deposited at 30 degrees of latitude will be 45 degrees.

magnetic measurements of rocks show that the fossil north pole was in the same place as the present one—called normal polarity—and in other cases the fossil south pole was located at the present north pole—called reverse polarity (see fig. 1–3). Thus, the history of the earth is divided into periods of normal and reversed polarity, which vary greatly in duration. The normal and reversed episodes can be recognized most perfectly in the cooled basalt lavas on the floors of oceans, but they are also recorded in many sequences of sedimentary strata on land. Because the earth's magnetic field applies all over the earth, each paleomagnetic episode is imprinted on all rocks that are being formed at the time of its occurrence. When sequences of strata with the same pattern of reversed and normal polarities are identified in different parts of the world, geologists know that they are of exactly the same age even though the sequences may be of different thicknesses in different regions (fig. 1–6). These discoveries were crucial to our understanding of how the plate tectonic model works.

The Plate Tectonic Model

The essence of the plate tectonic model is that the earth has a relatively rigid outer layer called the lithosphere that is about 75–125 kilometers thick (fig. 1–7). The lithosphere, which is capped by a thin crust beneath the oceans and a thicker continental crust elsewhere (see fig. 1–7), is broken up into large and small plates by the plastic flow of the hotter, denser rocks underneath, which, owing to the considerable heat generated in the interior of the earth by naturally radioactive minerals,

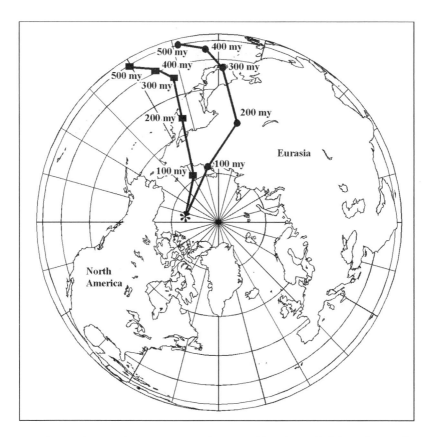

1–4. A map showing the various points (connected by a line called the polar wandering curve) at which paleomagnetic measurements predict that the pole would have been for each of the time periods indicated. Note that the rocks measured in Europe (black squares) give a different pole location for each time than the rocks measured in North America (black circles). Because we assume that there could only have been one north and one south magnetic pole at any one time, the two curves prove that each continent has moved in different directions. If the continents are moved to make their different polar wandering curves coalesce, Europe is placed next to North America and the Atlantic Ocean disappears. After "The Theory of Plate Tectonics," Copyright 1994, Tasa Graphic Arts, Inc.

are slowly churning. This deeper layer, 2900 kilometers thick, is referred to as the mantle, and it surrounds the even more dense core of the earth (see fig. 1–7). The flow of the mantle below the lithosphere follows a convecting pattern, so that in some zones in the earth hot liquid rock, called magma, rises to the surface and spreads outward and in others the cooler mantle is sinking (fig. 1–8). The crustal plates are riding passively on the back of this churning mantle. The interaction of

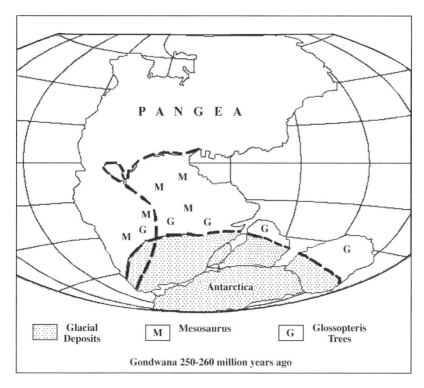

1–5. Using paleomagnetic evidence from rocks 250–260 million years old, geologists have reconstructed a map of the world for that time period, shown here. It shows a unified group of southern continents in one supercontinent named Pangea. The freshwater *Mesosaurus* fossils now cluster together in one region, and the *Glossopteris* trees form a latitudinal belt north of a unified southern polar ice cap. After "The Theory of Plate Tectonics," Copyright 1994, Tasa Graphic Arts, Inc.

these plates as they separate, collide, or pass by each other forms all the major geological features of the surface of the earth.

Perhaps the most striking feature of the crust is that it is divided into continents and oceans. To most people, this simply means land versus water. To geologists, the difference is much more fundamental (fig. 1–9). Continental crust is much thicker and on average less dense and structurally more complicated than oceanic crust. The continental crust is made of rocks that have been intensely deformed and chemically and physically altered, including some as old as 3 or 4 billion years. The ocean crust, on the other hand, is thinner, more dense, and relatively unaltered chemically and is nowhere more than about 200 million years old. Furthermore, the oceanic crust is less deformed and carries a record of the strength and polarity of the earth's magnetic field laid out in parallel stripes on either side of long central ridges that are

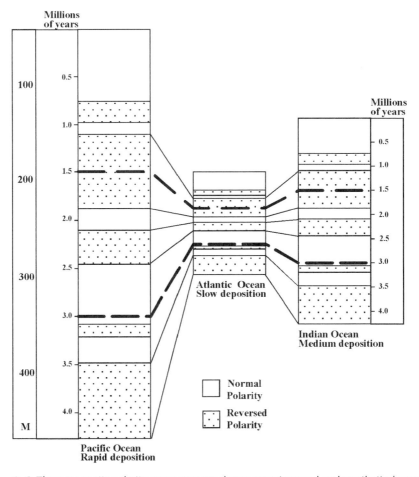

1–6. Three magnetic polarity sequences are shown superimposed on hypothetical cores of sedimentary sequences of the same general age that are of different thicknesses in the Pacific, Atlantic, and Indian Oceans. The sedimentary sections vary in thickness owing to different rates of deposition. In each core, however, the polarity sequence pattern is preserved identifiably, although one is proportionately expanded (rapid deposition) and another is proportionately contracted (slow deposition). The figure indicates lines that can be drawn to identify the same time in each section.

the sites of magma rising to the surface. Each polarity stripe on one side of the central ridge has a mirror image on the other side (fig. 1–10). The reasons for this will become clear below. These geologic differences between continental and oceanic crust are readily explained in the plate tectonic model by the different ways in which the various plates are formed and interact.

The interactions are of three main types. The first, in which two plates collide, is called a *subduction zone*. There are three possibilities.

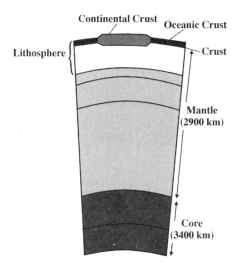

1–7. The diagram shows the thicknesses of the different layers that geologists use to define the internal structure of the earth. Note the contrast in thickness of the oceanic and continental crust.

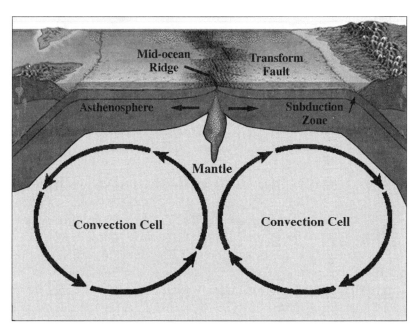

1–8. A schematic cross section of the earth showing the pattern of convecting magma in the mantle, the locations where magma rises to the surface, where the lithosphere is carried laterally, and where magma and the lithosphere sink. This mantle flow drives the movement of surface crustal plates. After "The Theory of Plate Tectonics," Copyright 1994, Tasa Graphic Arts, Inc.

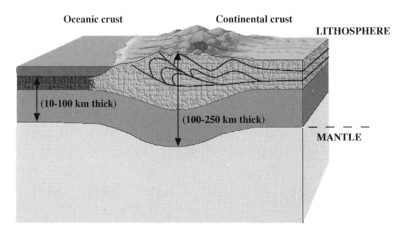

Oceanic crust   Continental crust

LITHOSPHERE

(10-100 km thick)

(100-250 km thick)

MANTLE

1–9. A schematic cross section of the solid lithosphere showing the differences in thickness and structure of oceanic versus continental crust. After "The Theory of Plate Tectonics," Copyright 1994, Tasa Graphic Arts, Inc.

The two plates may both consist of oceanic crust, or one may be oceanic and the other continental crust, or both may be continental, as in the Himalayas, where India and Asia have collided. In Central America, only the first two possibilities apply. In these collisions, the denser plate sinks beneath the other, usually causing a deep, narrow oceanic trench to form (fig. 1–11). The sinking plate starts to melt at a depth of about 70 kilometers, and the lighter elements of its magma rise to the surface through the overlying plate. If that plate is thin, as in oceanic crust, these magmas pour out as lavas or explode as mixtures of gas, ash, and molten rock fragments called *tephra,* along a curved line of volcanoes that form a volcanic island arc (fig. 1–11A). If the overlying plate is continental and thick, the magma often does not reach the surface and cools to form large bodies of granitelike rocks known as *batholiths* within the upper plate (fig. 1–11B). Eventually, erosion may expose these batholiths at the surface. The compression and buckling up of the overlying plate in subduction, the addition of these lighter magmas, and the scraping off of sediments and other rocks from the sinking plate onto the overriding plate all cause the crust of the upper plate to become thicker and thus more "continental" with time. As a result, the overlying plate subtly changes its chemical composition.

    The collisional plate junction, or subduction zone, is where continental crust is mainly generated. Once formed, this lighter, thicker crust will generally not be subducted but will continue to override oceanic crust; it thus continues to accrete material and grow steadily, differen-

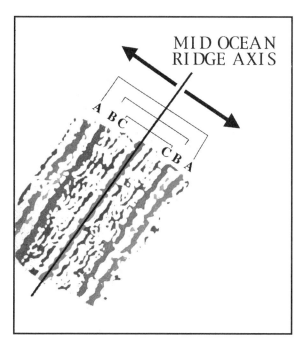

1–10. A magnetic map of the Atlantic sea floor to the southwest of Iceland shows the parallel stripes of normal and reversed polarity; each one to the east of the mid ocean ridge has a corresponding mirror image to the west as indicated at A, B, and C. The further the stripe is from the mid ocean ridge, the older it is. By dating the stripes radiometrically, via the geomagnetic reversal timescale, the rate of sea-floor spreading can be calculated. Modified from J. R. Heirtzler et al., "Magnetic Anomalies over the Reykanes Ridge," *Deep Sea Research* 13, no. 3 (1966): 427–443, fig. 1, and F. J. Vine, "Magnetic Anomalies Associated with Mid-Ocean Ridges," in R. A. Phinney, ed., *The History of the Earth's Crust, A Symposium,* 1968, fig. 6.

tiating into ever larger areas of thicker, lighter crust that now are called continents. Geologists believe that this process has been happening throughout the history of the earth. For this reason, continental crust has expanded in area and become more structurally complicated and more varied in composition than oceanic rocks. Modern subduction zones, involving continental crust, are the location of the world's mountain belts, deep oceanic trenches, zones of major earthquakes, and chains of explosive volcanoes. The same magmas that emplace the granitelike rocks carry with them the fluids from which precious metals are deposited and are thus often the sites of important mining regions, as is the case in much of Central America. Subduction zones in which two oceanic plates interact at first form only a volcanic island arc, such as the Lesser Antilles or the Aleutian Islands, although they are accom-

SOUTHERN CENTRAL AMERICA
(Early Stage of Panamanian Isthmus)

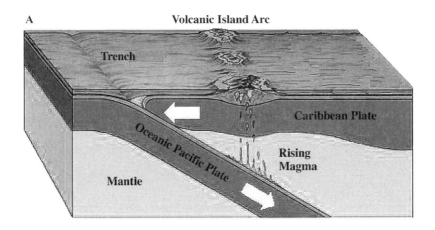

A — Volcanic Island Arc

Trench

Caribbean Plate

Oceanic Pacific Plate

Rising Magma

Mantle

NORTHERN CENTRAL AMERICA

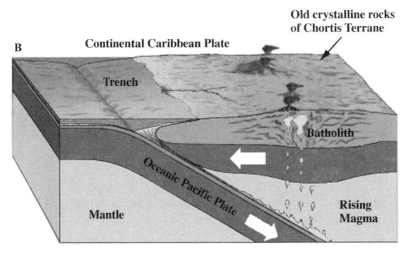

B — Continental Caribbean Plate

Old crystalline rocks of Chortis Terrane

Trench

Batholith

Oceanic Pacific Plate

Mantle

Rising Magma

1–11. Cross sections of subduction zones. In (A) each plate is oceanic crust. An island arc of volcanoes represents an early stage in the formation of southern Central America; (B) a subduction zone, where an oceanic crustal plate sinks beneath a continental crustal plate. This is a model for the structure of northern Central America. The batholiths and the surrounding metamorphic sediments, when eroded and exposed at the surface, will form the Central Crystalline Highlands. After "The Theory of Plate Tectonics," Copyright 1994, Tasa Graphic Arts, Inc.

panied by linear ocean trenches and major earthquake and volcanic activity (see fig. 1–11A). Later, they may form a continuous strip of land, as in the Central American isthmus.

The second type of plate interaction occurs where mantle magmas rise and pierce the lithosphere, creating new oceanic crust. This type may take two forms. In the first, the magma rises at a single point on the earth's surface so that a *hot spot* is formed and a very large volcano results, such as Kilauea in Hawaii. Enormous quantities of primal magma pour out. Geologists believe that because hot spots are direct emissions from the underlying mantle they are stationary with respect to the movements of the lithosphere. Thus, as the lithosphere moves over a hot spot, the volcano at the surface will move along the plate through time, leaving a linear track of extinct volcanoes behind in the same way a piece of paper moved steadily over a lit candle would leave a burned trace. The trace of the flame would mark the direction of the movement of the paper, and the track of the extinct volcanoes similarly marks the movement direction of the plate.

More commonly, rising mantle magma wells up along fissures that may be hundreds or thousands of kilometers long. Because hot, dense magma is rising, the surface of the crust is raised into a broad ridge, although the actual line along which the magma reaches the surface and spreads outward is a down-faulted valley or rift (fig. 1–12A). Because these rifted ridges occur in the center of oceans, this type of plate interaction is known as a *mid ocean ridge.* Mid ocean ridges allow geologists to get a glimpse of the rock (called *peridotite*) produced directly from the mantle, although generally the ocean crust forms by decanting of the peridotite magma into the mid ocean ridge, with rapid solidification into hard, black, dense basalt. All of the world's ocean floors are composed of basalt, which has neither mixed with other rocks nor fractionated and been purged of its lighter elements. This explains why the oceanic crust is thinner but denser than the continental crust differentiated in the subduction zones. The fractionated rocks that form continents have the general characteristics of granite and in origin and chemical composition strongly contrast with the typical basalt of oceanic crust. It follows from the differences between oceanic and continental crust that as mid ocean ridges form they will be the sites of new oceans, and as the new crust spreads out the ocean will grow in size (fig. 1–12B).

When the rising magma reaches the surface in the mid ocean ridge and begins to cool, the myriad tiny magnetite crystals within it align themselves with the earth's magnetic field at the angle corresponding to

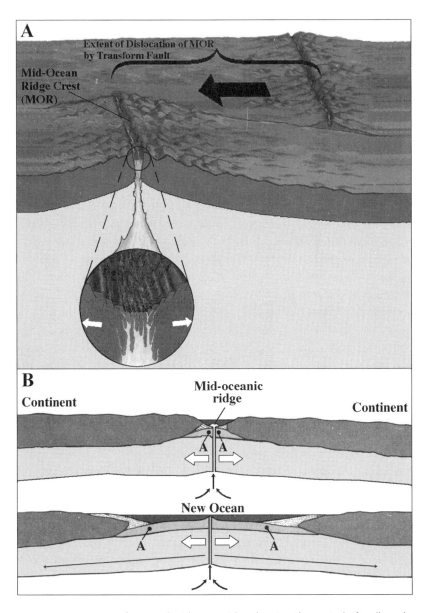

1–12. A cross section of a typical mid ocean ridge showing the central rift valley where magma (basalt) is extruded to form new oceanic crust: (A) the mid ocean ridge dislocated by a transform fault; (B) a mid ocean ridge slowly creates a new ocean underlain by young oceanic crust. After "The Theory of Plate Tectonics," Copyright 1994, Tasa Graphic Arts, Inc., and Harold Levin, *Contemporary Physical Geology* (Saunders College Publishing Co.), 271, fig. 10–14.

the earth's latitude at that point. They also take on the polarity and the intensity of the earth's magnetic field at the time of cooling. Once the basalt is cooled and hard, the magnetic signal is frozen in place. As the oceanic crust spreads outward in opposite directions on either side of the mid ocean ridge, new magma wells up to take its place. If the earth's magnetic field has changed polarity and intensity during this process, then the new basalt will differ in paleomagnetic signal from the previously formed basalt, which now forms two mirror-imaged stripes on the ocean floor; located on either side of the newly emplaced basalt, the two stripes are identical in polarity, magnetic intensity, and age (see fig. 1–10). As the process continues over millions of years, new mirror-imaged stripes are continuously formed on either side of the mid ocean ridge, and the older ones migrate farther and farther apart. Each stripe, running the length of the opening ocean, has the same age and polarity everywhere. Geologists can establish the age of the stripe by analyzing the amount of decay of radioactive minerals within a piece of basalt at one point, a process called radiometric dating, and thus establishing the age of that stripe anywhere along its length. If the age of the stripe and the distance from the center of the mid ocean ridge are known, then the speed at which the basalt floor is spreading out from the mid ocean ridge can be calculated. About 7 to 8 centimeters per year is a typical rate for these movements. At its current rate of spreading, 2.5 centimeters per year, the sea floor of the Atlantic Ocean has taken 230 million years to reach its present width!

The oceans, then, through their mid ocean ridges, are where new primal crust originates, and this crust, carried outward in both directions, bears a virtual tape recording of the age, width, and paleomagnetic history of the ocean at each stage of its growth. As the same process is taking place in two or three oceans simultaneously, the phased sequences of polarity reversals will enable geologists to correlate stripes formed at the same time in each ocean. Moreover, sediment accumulated in vertical sequences on other parts of the ocean floor and on land records the same episodes of normal and reversed polarity (see fig. 1–6). Within 200 to 300 million years, the spreading ocean floor will eventually arrive at a subduction zone, where it will sink and be melted. For this reason there is no ancient oceanic crust left on earth.

The third and last type of plate interaction is known as a *transform fault*. Plates cannot move away from all mid ocean ridge spreading centers at the same velocity because of the spherical shape of the earth and because lava is generated at varying rates in different places. In order for portions of the crust to move at different velocities, they have to break into segments along faults that transform the motion of the spreading crust into units (plates) with different velocities—hence the term transform fault. Unlike the other two, this interaction does not involve vertical movements and thus little or no volcanic activity and no destruction or creation of crustal material. In transform faults, plates slide by each other, dislocating previous geologic structures (see fig. 1–12A). The most distinguishing feature of a transform fault is intense earthquake activity. Transform faults offset preexisting subduction zones and mid ocean ridges so that the earth's crust is now a mosaic of small and large plates, all of which are interconnected by a web of these three junctions (fig. 1–13).

The plate tectonic model has for the first time allowed geologists to unite, in a single unified theory, the previously unconnected patterns of mountain ranges, volcanic activity, earthquakes, and oceanic trenches, as well as the otherwise inexplicable patterns of fossil and rock distributions in the geologic past. I shall use it to unravel the recent geological history of Central America, which has been dominated by the interactions of a major subduction zone running along its Pacific margin and two transform faults bounding it to the north and south. Central America is affected by the interplay among five different plates, and within its boundaries there are two triple junctions—very complicated points at which three plates intersect. For these reasons its geological history has been complicated and violent.

Assembling Central America

Modern Central America has been geologically recognizable only in the past few million years. Before that, the geological units that form the present isthmus either had not yet formed or were located at other latitudes. A geological unit that has originated in one location and by plate movement has been transported, often large distances, and then accreted onto the edge of another plate during subduction is known as an *exotic terrane*. Several such terranes are now closely united to form modern Central America, but they evolved in very different environments and hence have strikingly diverse rocks and fossils, which testifies to their separate history and genesis. The geologic term *terrane* is distinct from *terrain,* which refers to the nature of the land surface.

Geologically, Central America has been strongly affected by the movements of South and North America. Two hundred fifty million years ago, these two continents were part of the great supercontinent Pangea. About 140 million years ago, at the end of the Jurassic period,

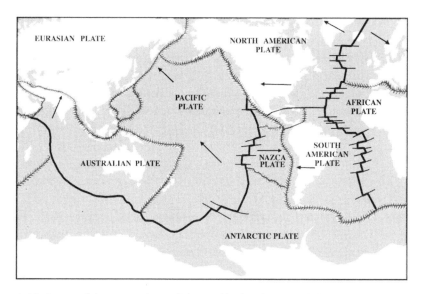

1–13. A map of the major plates of the world. The plates are defined by connecting the mid ocean ridges and transform faults (thick black lines), and the subduction zones (hachured black lines). After "The Theory of Plate Tectonics," Copyright 1994, Tasa Graphic Arts, Inc.

Pangea, through the formation of a mid ocean ridge, began slowly to rift apart. At first, North America separated from Europe, North Africa, and South America to form the fledgling Atlantic Ocean, which connected directly to the Pacific through the present location of Central America (fig. 1–14). On the southern margin of the North American plate, the future Mexico formed a peninsula, to which were attached the Maya and Chortis terranes. Here, more than 300 million years ago, the sediments that would come to form the Crystalline Highlands of northern Central America were deposited. The Maya Terrane was to stay locked to Mexico in a stable position. By contrast, the Chortis Terrane would become detached from western Mexico.

By 80 million years ago, late in the Cretaceous period, the widening Atlantic Ocean had now spread southward and was separating Africa and South America (fig. 1–15). At the same time, the Galápagos hot spot started a vast outpouring of basalt that covered an area 1000 by 3000 kilometers (the gray area in figure 1–15). The Pacific Ocean now had its own mid ocean ridge system, dividing an eastern Farallón Plate from a western Pacific Plate. The eastern margin of the Farallón Plate was now a subduction zone with an accompanying active volcanic arc that stretched along the western coast of North America, including the Chortis Terrane at its southern tip, and the island arc across the site of

Central America, and down the west coast of South America (see fig. 1–15). The island arc in the location of Central America is labeled *GA* on the figure because subsequent plate movements would carry it far to the northeast, where it would eventually form the islands of Jamaica, Cuba, and Hispaniola (that is, the Greater Antilles).

By the end of the Cretaceous period, about 65 million years ago, the great volcanic arc along the west coast of all the Americas was ruptured in the region of Central America, and a segment of the arc, together with the great basalt sheet poured out by the Galápagos hot spot, had been squirted northeastward as a new small unit—the Caribbean Plate (fig. 1–16). On its western margin a new subduction zone and volcanic island arc had formed—the geological beginnings of modern Central America. The great Galápagos "flood" basalt now floors the Caribbean Sea, and fragments of it in turn have been scraped off in subduction movements and preserved in many terranes that rim the Caribbean.

The future Central America at the end of the Cretaceous period

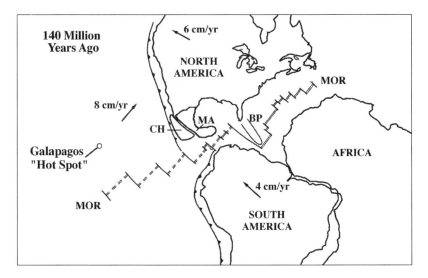

1–14. A reconstruction of the arrangement of plates 140 million years ago, at the end of the Jurassic period. Arrows indicate the direction and speed of relative movement of the plates. Note that only the North Atlantic has opened, connecting with the Pacific Ocean, and that the Chortis Terrane lies largely to the west of the Maya Terrane. BP = Bahama Platform; CH = Chortis Terrane; MA = Maya Terrane; line with black triangles is a subduction zone; double line with offsets is a mid ocean ridge (MOR) with transform faults. (Symbols are the same for figures 1–15, 1–16, 1–17, 1–18, and 1–20.) Modified from Duncan and Hargraves, in W. Bonini, R. B. Hargraves, and R. Shagam, eds., *The Caribbean South American Plate Boundary and Regional Tectonics,* Geological Society of America Memoir 162, fig. 1.

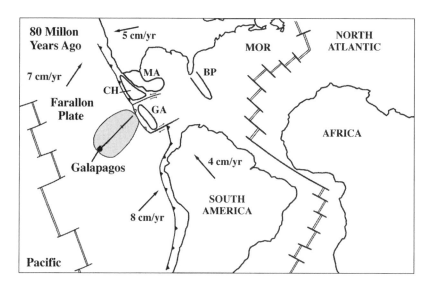

1–15. A plate reconstruction for 80 million years ago, late in the Cretaceous period. The shaded area is a "flood" basalt from the Galápagos Hot Spot, and GA represents a volcanic arc destined to become the future Greater Antilles. Modified from Duncan and Hargraves, in W. Bonini, R. B. Hargraves, and R. Shagam, eds., *The Caribbean South American Plate Boundary and Regional Tectonics,* Geological Society of America Memoir 162, fig. 4.

consisted, then, of a southern extension of the North American Plate forming a continental peninsula, present-day Mexico, with an easterly extension formed by the Maya Terrane, presently underlying the Yucatán, El Petén, and Belize. Sutured onto the Maya Terrane to the south was the Chortis Terrane, newly arrived from the northwest, which now underlies El Salvador, southern Guatemala, Honduras, and Nicaragua (fig. 1–16). The suture between the Chortis and Maya terranes runs along the Motagua Valley in Guatemala, where contemporaneous rocks completely different in character and in original latitudes face each other on opposite sides of the valley. The future Costa Rica and western Panama were a series of oceanic volcanic islands stretching to the south of the Chortis Terrane (labeled *CA* in figure 1–16).

During the Eocene period, about 40 million years ago, the volcanic arc bordering the Caribbean Plate to the northeast finally collided with the Bahamas-Florida Platform, effectively preventing it from moving further northeastward. Cuba and Hispaniola were formed as a result of this collision. After this time, the Caribbean Plate began to move eastward and a new subduction zone was formed, now located along the Lesser Antilles Volcanic Arc (fig. 1–17).

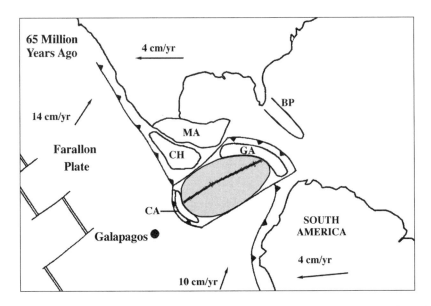

1–16. A plate reconstruction for 65 million years ago at the Cretaceous-Tertiary boundary. Note that the Chortis Block now lies to the south of the Maya Terrane and is fused to it; the Greater Antilles arc has moved far to the northeast, and the new Caribbean Plate (shaded area), floored with the Galápagos "flood basalt," is delimited to the west by a new subduction zone forming the Central American volcanic arc (CA). Modified from Duncan and Hargraves, in W. Bonini, R. B. Hargraves, B. and R. Shagam, eds., *The Caribbean South American Plate Boundary and Regional Tectonics,* Geological Society of America Memoir 162, fig. 5.

Meanwhile, the Farallón Plate in the north had been entirely subducted under the North American Plate, and the Pacific mid ocean ridge system now intersected the coast near the Mexico–United States border. Farther south, the Farallón Plate continued to converge on the Chortis Terrane as well as the proto–Central American arc and South America (see fig. 1–17).

By the beginning of the Miocene period, 20 million years ago, the Caribbean Plate had extended considerably to the east (fig. 1–18). An oceanic gap between the Central American volcanic arc and South America had developed, serving to keep the terrestrial faunas of North and South America separated. In addition, the Farallón Plate had now separated into two units, a northern Cocos Plate and a southern Nazca Plate. Figure 1–19A is a tentative reconstruction of the geography of the isthmus about this time.

As a result of these plate movements, a gradual closing of the deep-water connection of the Pacific and Caribbean started at this time. The proto–Central American volcanic arc extended eastward, and about 12

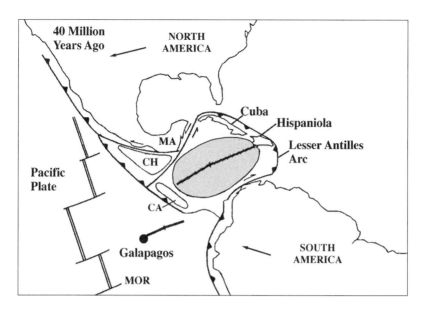

1–17. A plate reconstruction for 40 million years ago, late in the Eocene period. The Greater Antilles volcanic arc has now collided with the Bahama Platform, and the new movement of the Caribbean Plate is eastward along the Lesser Antilles arc. The Americas have overrun the Pacific mid ocean ridge system and much of the Farallón Plate; there is a continuously active volcanic arc the length of the eastern Pacific. Modified from Duncan and Hargraves, in W. Bonini, R. B. Hargraves, and R. Shagam, eds., *The Caribbean South American Plate Boundary and Regional Tectonics,* Geological Society of America Memoir 162, fig. 6.

million years ago finally collided with South America. The future isthmus rose to form a sill some 1000 meters deep.

This momentous event triggered profound changes in both oceans that are still going on today. The first result of the severing of deep-water (2000 meter) connections between the oceans was the disappearance in the Caribbean of microscopic plants (*diatoms*) and animals (*radiolaria*) whose skeletons are made of glass, or silica. In the Pacific, however, this important component of the plankton continues to be abundant to the present time.

By 11 million years ago, islands may have begun to appear in the present location of eastern Panama and the southern half of modern Central America. Over the next few million years the region became an archipelago with many varied marine and coastal habitats. This archipelago further restricted the marine circulation from the Caribbean to the Pacific, but other factors now began to play a role also.

The Antarctic ice cap had began to grow so that sea level dropped as ice was sequestered in the polar caps, and sea temperatures cooled.

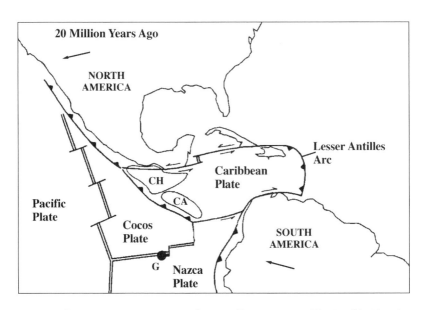

1–18. A plate tectonic reconstruction for 20 million years ago. The Farallón Plate is now split into the Cocos and Nazca plates, and the Caribbean Plate continues to migrate eastward, at the same time being squeezed by the relative northwestward movement of South America. Modified from Duncan and Hargraves, in W. Bonini, R. B. Hargraves, and R. Shagam, eds., *The Caribbean South American Plate Boundary and Regional Tectonics,* Geological Society of America Memoir 162, fig. 7.2.

Some geologists believe that this brought the cool, southerly flowing California Current, which is presently restricted to the north, as far south as Guayaquil in Ecuador, effectively isolating the marine species living in the Caribbean from those in the Pacific. Microscopic single-celled animals with calcareous skeletons called *foraminifera* are very distinctive of sediments at the bottom of the sea, different species living at various depths and in diverse sediment conditions. Those typical of the present California Current are known from 11 to about 6 million years ago in sediments along the Pacific coast as far south as Guayaquil. Caribbean forms are not mixed with them. While the marine species were apparently strongly separated at this time, a few island-hopping, swimming animals such as raccoons and sloths were able to migrate between North and South America (see chapter 4) as the isthmus steadily rose and more islands appeared. About 6 million years ago the isthmian sill would have been only 150 meters deep (fig. 1–19B).

Between 6 and 3 million years ago, further dramatic regional changes occurred, culminating in the final closure of the land barrier between the Pacific and the Caribbean. But first, the California Current seems to have retreated northward again so that once more marine

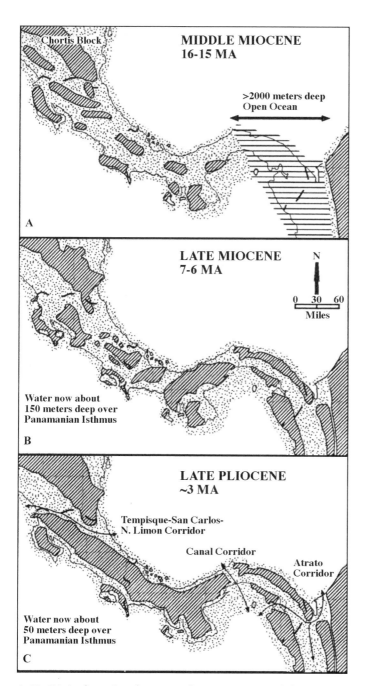

1–19. (A) A schematic paleogeographic interpretation of the Central American isthmus 15 million years ago, in the middle of the Miocene period. The dotted pattern indicates the approximate position of the marine shelf; (B) a schematic paleogeographic interpretation of the Central American isthmus 6 million years ago near the end of the Miocene period. The deepest part of the marine shelf along the isthmus is now about 150 meters; (C) a schematic paleogeographic interpretation of the Central American isthmus about 3 million years ago, at the end of the Pliocene period. The probable last marine corridors connecting the Pacific to the Atlantic are indicated.

species from the Caribbean and the Pacific intermingle. Because of the increasing uplift of the massifs in the San Blas and Majé areas of Panama and in the neighboring South American Andes, huge amounts of sediment, several thousands of meters thick, were eroded from their flanks and rapidly filled up the basins between the rising islands and peninsulas. Thus, about 4 million years ago the deepest water along the isthmian archipelago might have been only 50 meters deep (fig. 1–19C), and after about 3 million years the isthmian barrier was complete. This closing enabled a flood of terrestrial animals to cross the new land bridge between North and South America, as described in chapter 4. The structure and plate movements of the Central American region today are summarized in figure 1–20. The figure shows that the collision of South America with both the Caribbean Plate and the Cocos Plate has produced new subduction zones or overthrusts: one north of the Caribbean coasts of Colombia and Venezuela and one north of Costa Rica and Panama that is buckling up and bending southern Central America.

Southern Central America

The foregoing narrative helps to explain the striking differences of topography and geology between north and south Central America. Figure 1–19C shows the last probable marine corridors between the Pacific and the Caribbean just before final closure. The final strait between South and Central America may have been in the region of the present Atrato Valley of Colombia and the Tuira and Chucunaque valleys of the Darién in Panama. Compression and uplift of this region have only recently brought sediments typical of deep oceans to the surface. The higher areas—for example, the massifs of San Blas, Majé, and Sapo (see fig. 3–4)—that formed the first rising islands and ridges during the early part of the isthmian collision appear to mark the locations of former subduction zones. The intense buckling and resulting compressional thickening, however, as well as a complication in the plate movements to be explained later, have meant a general absence of volcanoes in the Darién of Panama during its later geological history. Today, the great Central American volcanic arc that stretches from Mexico throughout Central America terminates in the apparently quiescent volcano of El Valle in central Panama, although there are very recent lava flows near Panama City.

The volcanic spine of Central America is certainly its most striking physical feature and accounts for the violent explosive activities of

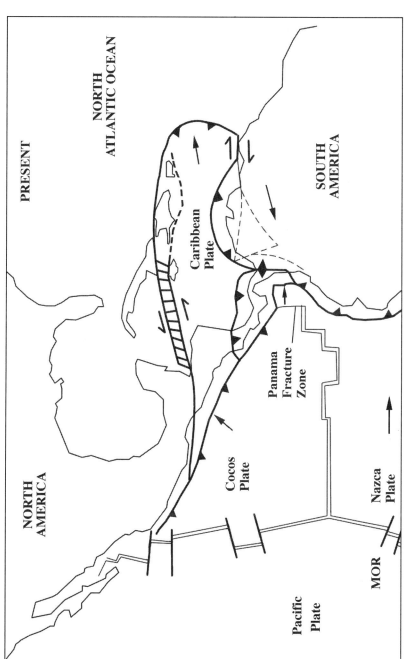

1–20. The plate configurations of the Caribbean and Central American region as they are today. Note the subduction segments north of South America and north of southern Central America, evidence of the collision of the South American and Caribbean plates and the reason for the uplift of the Isthmus of Panama. Modified from Duncan and Hargraves, in W. Bonini, R. B. Hargraves, and R. Shagam, eds., *The Caribbean South American Plate Boundary and Regional Tectonics*, Geological Society of America Memoir 162, fig. 8.

such famous peaks as Fuego, Izalco, Masaya, Arenal, Poas, and many others. As noted above, these volcanoes are the manifestations of the fractionation of the lighter molten elements from the Pacific Plate as it sinks beneath the Caribbean Plate. These powerful compressive and buckling forces have bent the Pacific crust down into the long Middle America Trench, more than 5000 meters deep, and are the cause of the catastrophic, ubiquitous earthquakes that plague Central America from western Panama to Guatemala.

From central Panama to northern Costa Rica, the Central American isthmus continues to be relatively narrow and dominated entirely by a volcanic arc made up primarily of lavas and rocks formed from volcanic eruptions. Marine sediments are found mostly in marginal basins and on the marine shelves flanking the chain. Some of these, including the basins of Bocas del Toro in Panama and the basins of Limón and Terraba in Costa Rica, have been uplifted and incorporated onto the isthmus, (see fig. 3–4); these basins contain rich deposits of marine fossils that allow the various stages in the uplift of the isthmus to be reconstructed.

Exotic Terranes

Two other striking geological events have diversified the topography of the lower isthmian volcanic chain in Costa Rica and Panama. First, the continual convergence of the Pacific Plate beneath Central America has swept a variety of *exotic terranes,* such as large hot-spot volcanoes, oceanic ridges that are the traces of hot spots, and ancient mid oceanic ridges, into the mouth of the Central American subduction zone. Because of their thickness and slightly lighter density, the terranes do not subduct easily and so become accreted onto the trailing edge of the Caribbean Plate, where they manifest themselves as promontories: among these are the Azuero, Burica, Osa, and Nicoya peninsulas (fig. 1–21). Each of these regions has at its core rocks that were originally formed perhaps thousands of kilometers to the south and west. Through measurement of the paleomagnetic orientation of the magnetite crystals in rocks from these terranes, the original locations of these far-traveled geological elements are now being established.

The Cocos and Nazca Plates

The second disruptive geological event in the history of the lower Central American volcanic chain and subduction zone requires a close look at the details of the complicated plate movements to the south of Costa

Rica and Panama. When the Farallón Plate split into the Nazca and Cocos plates, the junction between them was a mid ocean ridge system (see fig. 1–20). At its eastern margin, however, the junction is a transform fault known as the Panama Fracture Zone, which intersects the isthmus close to the Burica Peninsula, at the border of Panama and Costa Rica (figs. 1–20, 21)

Relative to the Caribbean Plate to the north, the Cocos and Nazca plates move in different directions. The Cocos Plate moves at about 8 centimeters per year to the northeast almost at right angles to the margin of the Caribbean Plate, with the result that subduction is vigorous and volcanoes and earthquakes have always been active, as they still are today. To the east of the Panama Fracture Zone, however, the Nazca Plate appears to move eastward so that for most of the length of Panama there is little difference in direction or speed of movement between the Caribbean and Nazca plates (see fig. 1–20). This may explain the relative quiescence of volcanic activity in eastern Panama and the absence of frequent severe earthquakes. Volcanic activity in Panama is not completely dormant, however, for there is good evidence that there have been lava flows and perhaps some eruptions between Volcán Barú and El Valle in the past few hundred years.

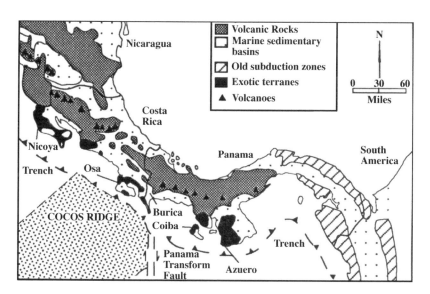

1–21. A geologic map of Central America showing the location of Exotic Terranes and the Cocos Ridge. The Panama fracture zone is a transform fault that separates the Cocos and Nazca plates.

The Cocos Ridge

Also intersecting the isthmus in the region of the Burica Peninsula, but lying to the west of the Panama Fracture Zone, is the Cocos Ridge (see fig. 1–21). It consists of a welt of lighter oceanic crust 2000 meters high and 200 kilometers wide, apparently representing the trace on the Cocos Plate of its passage over the hot spot located at the Galápagos Islands. For the past 3 million years it has apparently been subducted with great difficulty owing to its extra thickness and lightness, and this is having a spectacular effect on the Central American volcanic arc and subduction zone. First, it has uplifted and indented the Middle America Trench, creating a difference in sea floor elevation of more than 3500 meters. Second, it has apparently raised the isthmus to form the Cordillera de Talamanca, at almost 4000 meters, one of the highest points in Central America. Across the Talamanca range all volcanic activity has been choked off, so that while there is normally a volcano every 28–30 kilometers along the isthmian volcanic chain, in the Talamancas there are none for 125 kilometers. Indeed, the whole of the volcanic chain is domed up from El Valle, Panama, in the east (1200 meters) to Arenal, Costa Rica, in the west (1600 meters), a distance of 570 kilometers (fig. 1–22).

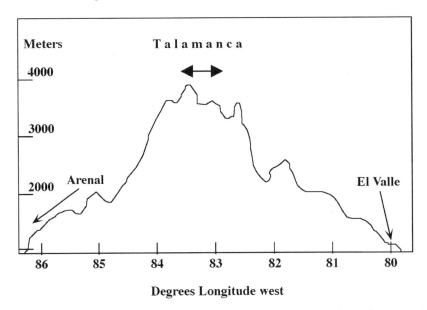

1–22. A topographic profile along the southern Central American isthmus from Arenal in northwestern Costa Rica to El Valle in central Panama. The topographic elevation reflects the uplift of the isthmus imposed by the insertion of the Cocos Ridge beneath it in the region of the Talamanca Mountains.

Northern Central America

Northwest along Central America into Nicaragua, the isthmus becomes wider and the age and complexity of the geology correspondingly greater. All of the crust that forms southern Central America, from southernmost Nicaragua to the Darién, has evolved from volcanic activity that generated exclusively oceanic and relatively young crust. To the north, the geologic history centers around the Chortis and Maya terranes, two much larger, older, and more complex regions than any that occur elsewhere in southern Central America.

The Chortis Terrane

The Chortis Terrane is first encountered in central and northern Nicaragua, north of the Nicaragua Depression, and it underlies El Salvador, Honduras, and Guatemala south of the Motagua Valley. The Pacific coastal margin of this terrane is largely dominated by geologically recent lavas and associated volcanic rocks (fewer than 2 million years old) that blanket the region from the coast inland for at least 100 kilometers (see figs. 3–1, 3–3, 3–4) and are a continuation of the volcanic arc that created southern Central America. These volcanoes are still so active that they have poured out a staggering 16 cubic kilometers of lavas and ash along a distance of 1100 kilometers since A.D. 1680.

Although still a striking physiographic feature, the volcanic chain is only one of five distinct geologic regions comprising the Chortis Terrane, several of which long predate the beginning of the Central American volcanic arc.

The other regions making up the Chortis Terrane are the Northern Sierras of southeastern Guatemala and northern Honduras, which represent the intense shearing zone where the northern margin of the Caribbean Plate (the Chortis Terrane) is now moving eastward relative to the southern limit of the North American Plate (the Maya Terrane); the Central Crystalline Highlands of Guatemala and Honduras (see figs. 3–1, 3–3), which contain the most ancient rocks in Central America; the High Volcanic Plateaus, a distinctive region of highlands running inland of the modern volcanic arc from southern Guatemala through El Salvador and Honduras to Nicaragua (see fig. 3–3), formed by volcanic activity 10 to 20 million years ago; and the huge Mosquitia embayment, occupying eastern Nicaragua and the extreme eastern portion of Honduras (see fig. 3–3). Each of these regions has a very different geological history, and they largely correspond to the distinctive physiographic regions of northern Central America described in chapter 3.

The Northern Sierras

The Northern Sierras, which stretch from northeast of Guatemala City in southern Guatemala along the northern coast of Honduras, are a series of east-west trending mountains formed of metamorphic and igneous (formed from molten magma) rocks, partly mantled by younger limestones. The orientation of these mountains is strongly controlled by the powerful and active shearing faults that now delimit the Caribbean and North American plates.

Across these valleys the ancient Maya and Chortis terranes are now facing each other, having become intensely deformed and altered during the plate movements that brought them together at the end of the Cretaceous period. Subsequent eastward movement of the Caribbean Plate, starting at least 40 million years ago, has now caused shearing between the two terranes. Some geologists think that the shearing has caused the two plates to be offset about 120 kilometers; others postulate more than 1000 kilometers of displacement.

The Central Crystalline Highlands

The Central Crystalline Highlands of Honduras form a rugged mountainous core to the Chortis Terrane, and ancient continental basement rocks dating back as far as 500 million years are at their root. These rocks have been subject to intense heat and pressure deep in the crust. Subsequent prolonged erosion finally exposed them so that by 140 million years ago they underlay a shallow marine shelf somewhere on the western margin of the separating and newly formed Atlantic Ocean. Unlike the Maya Terrane, Chortis was migrating south from western Mexico at this time, and its precise location is not known. On its shelf developed a thick sequence of limestones and reefs somewhat like the Bahamas today. Later, the marine shelf was uplifted and began to erode so that rivers and estuaries covered the limestone with gravel and red sand; periodically it would sink beneath the sea, and new reefs and lagoons would form. It then became broken up into fault-fractured valleys, called grabens, and intervening mountains (horsts) and must have looked much like the Basin and Range of Nevada today.

About 90 million years ago, the region was again flooded by the sea, an episode that also flooded large areas of North and South America. Once more a Bahama-like limestone platform was created. The faulted and uplifted elements of these limestones outcrop extensively in the Northern Sierras. From about 75 million years ago, the whole region was penetrated by molten magmas and intensely deformed and uplifted into the highland topography seen today. These granitic and metamorphic rocks underlie the limestones in the Northern Sierras but are extensively revealed in central Honduras and northern Nicaragua, where they give the Crystalline Highlands their name. These intense movements were part of a regional geological event in which Chortis became fused or sutured onto the Maya Terrane (see fig. 1–16). At the same time, the Central American volcanic arc continued subduction and volcanism along the length of the boundary between the Pacific and Caribbean plates, the present sites of western Guatemala, El Salvador, western Nicaragua, Costa Rica, and western Panama.

The High Volcanic Plateaus

The High Volcanic Plateaus are formed by an extraordinary pile of explosive volcanic debris that covers much of southern Guatemala, western Honduras, northern El Salvador, and west-central Nicaragua (see fig. 3–3). They also cap some of the mountains of the western part of the Crystalline Highlands of Honduras. Although there had been continuous if sporadic volcanic activity in this region because of the subduction of the Farallón Plate from the west since the Late Cretaceous more than 80 million years before, there occurred about 20 to 14 million years ago a truly stupendous outburst of volcanic eruptions that covered more than 10,000 square kilometers of Central America. Thousands of cubic kilometers of volcanic deposits were produced, varying in thickness from 700 to more than 2000 meters. During this phase, a special kind of eruption frequently happened wherein molten magma was choked or blocked in the vent of the volcano until it exploded. Gasses not able to be released until the explosion took place (as when a bottle of champagne is shaken before the cork is released) expanded violently, pulverizing the molten magma so that it erupted as glowing hot clouds of foaming gas, ash, and tephra that spewed over the landscape for hundreds of kilometers. These deposits, now faulted and dissected by erosion, form a series of high plateaus that are often capped by younger, more familiar basalt lava flows. On this older volcanic edifice has been built the spectacular line of recent and currently active volcanoes that form the modern Central American volcanic arc in this region.

The Mosquitia

The Mosquitia embayment (see fig. 3–3) in eastern Honduras and central and eastern Nicaragua today is a vast forested lowland, but for much of the past 100 million years it was mountainous. Although probably lying on an as-yet-undetected ancient continental crustal basement, 4500 meters of sedimentary rocks underlay this region, revealed mostly in boreholes. Until 35 million years ago, these deposits indicate a rugged mountainous land mass with active erosion of sediments into intermontane basins, terrain very different from that of the present day. By 35 million years ago the region had become worn down, so that it sank beneath the sea, and marine silt and limestone accumulated. The region rose above sea level again about 10 to 5 million years ago and ever since has oscillated between a shallow, swampy coastal shelf and an emergent coast with extensive estuaries and marginal sea grass beds and coral reefs much like those of today. Even slight changes in sea level now would flood much of this low-lying region.

The Maya Terrane

The most northerly part of Central America lies entirely on the Maya Terrane, part of the North American Plate that underlies Belize and the El Quiché, Alta Verapaz, and El Petén provinces of Guatemala. The Atlantic Ocean began to open about 230 million years ago, and the Maya Terrane formed part of its western continental shelf. Much of the earlier geological record is obscure, but the opening of the Atlantic Ocean was accompanied by geological extension of the crust, which produced fault-bounded rift valleys and intervening mountains, or horsts, along its margin. Erosion of the horsts filled the valleys with gravel and alluvium, and periodic evaporation formed salt and gypsum. These deposits, patchily distributed but distinctive in Central America and Mexico because of their rust-red color, show the Maya Terrane to have been emergent until about 150 million years ago.

For the next 90 million years, the Maya Terrane was submerged, becoming part of a passive (as opposed to subducting) Atlantic continental shelf along which geological conditions were stable and quiet. During this immense amount of time, an enormous reef and lagoon system evolved that generated vast quantities of limestone as well as evaporite deposits such as salt and gypsum. This is the origin of the 3000-meter-thick blanket of limestone that now covers much of the Petén, Belize, and, farther north, the Yucatán.

The limestone cover is largely absent in only two regions of the Maya Terrane (see fig. 3–1). On its southern rim, in the 3000-meter-high ranges that form the Sierras de Chuacús and Las Minas, between the Motagua and Polochic rivers, ancient continental crust is preserved, now intensely folded, sheared, and mineralized as a consequence of the collision of the Maya and Chortis terranes. At this time also, pieces of an old mid oceanic ridge that formerly lay between the two terranes and that contains samples of the mantle (peridotite) and ocean-floor crust were scraped off during subduction and then squeezed up into this zone where they are now exposed as intrusions into the older crustal rocks. This is the source of much of the jade carved in southern Central America by indigenous peoples (see chapter 6).

The second major breach of the great limestone plateau is in the Maya Mountains of Belize (see fig. 3–1). Here, in a 50-by-90-kilometer window, the 340-million-year-old crust that forms the granite basement of the Maya Terrane has been exposed by uplift and erosion of the overlying limestones.

The collision of the Maya and Chortis terranes at the end of the Cretaceous period signaled the end of the long period of reef growth, as the great limestone bank was raised out of the sea. Limestone terrains are highly susceptible to weathering owing to solution of the rock by rainwater, a process that produces a distinctive topography of circular and vertical-sided limestone towers and intervening basins called karst, as well as immense underground caverns and a general absence of surface drainage. These features give the Petén and much of Belize (as well as the Yucatán) their unique regional character within Central America, as described in chapter 3.

Closing of the Isthmus and the Ice Age

One of the striking consequences of the formation of the Central American isthmus is that the oceans on either side became different. These effects began 15 million years ago, as noted above, when the deep-water circulation between the Pacific and the Atlantic oceans began to be affected and plankton with silica skeletons disappeared from the Caribbean. From 10 to 5 million years ago, an extensive archipelago existed throughout the present region of Central America, forming a more complex and varied marine ecosystem than exists today along the two coasts of the isthmus. From 5 to 3 million years ago, the marine connections across the isthmus would have been narrow and meandering and were probably located in three areas (see fig. 1–19C). First, the Atrato Valley and the Gulf of Urabá were still connected through the San Juan River in Colombia, and the Tuira-Chucunaque rivers in the Darién to the Pacific Ocean. Second, a marine embayment may have connected the Caribbean via the Nicaragua Depression to the Pacific, and third, at least in the early part of this period there are likely to have been connections through the Chagres Valley along the present track of the Panama Canal. The questions of how the barrier of the isthmus was finally completed and whether it was subsequently breached are complicated by the fact that a different set of factors involving global climatic and sea level changes began to play an important role about this time.

The Effects of the Ice Age

About 3 million years ago, as the isthmus rose to become a shallow barrier forming an extensive archipelago of islands, the Ice Age began to develop as repeated phases of glaciation interspersed by warmer interglacials, in one of which we are now living. There is no reason to suppose that many more such glacial episodes will not come in the future. Starting about 2.5 million years ago, these cold phases became more and more pronounced and showed a remarkably constant frequency of about 100,000 years. In each glacial episode, temperatures got steadily colder and ice accumulated in polar ice caps, causing a fall in sea level. But each time at a given threshold, the process was reversed: temperatures rapidly warmed, the ice melted, and sea level quickly rose. The frequency of these oscillations corresponds very closely to predicted variations in heat coming to the earth from the sun as the distance and orientation of the earth changed during its orbit around the sun. These were predicted by a Yugoslav mathematician earlier in this century and are now called Milankovitch cycles after him. They resulted in changes in sea level that may have been as great as 180 meters, and researchers know that in the past 20,000 years sea level has risen about 135 meters as the modern glaciers have been melting. Thus, the isthmian barrier may have closed and then later been breached during one of these sea level rises. Because the lowest relief of the isthmus is only 45 meters along the Nicaragua–Costa Rica border, further sea level rise as the remaining glaciers melted could still almost breach the isthmus again.

Calculating Paleotemperatures

How do geologists know that temperatures seesawed as predicted by the Milankovitch cycles, and do they have any direct evidence for high and low sea levels? Two lines of evidence strongly point to these conclusions. First, reef corals grow only close to sea level, and by locating and dating corals of this type that are now many tens of meters below the present sea level geologists can calculate the degree of sea level lowering for different times in the past. When a historical sea level curve is constructed using this technique its fluctuations correlate in frequency to the oscillating Milankovitch cycles.

Second, shelled marine animals take up calcium, carbon, and oxygen from the seawater to make their shells. The element oxygen possesses different isotopes, variants of the element that have slightly different atomic nuclei and hence different properties. When certain marine animals secrete their shells, the ratios of oxygen isotopes in the shells change according to the temperature of the seawater. Thus, each shell is a recording thermometer for the temperature of the seawater in which it lived. When paleontologists find, at different stratigraphic levels, well preserved fossil calcareous shells that lived in the floating plankton or in the mud on the bottom of the sea, they can trace the changes in the temperature of the surface and bottom waters of the oceans for different times in the past by carefully measuring the ratio of the oxygen isotopes in the fossil shells. Paleontologists can thus track changes in marine climate as it responds to the glacial cycles, and the clearly oscillating pattern of Milankovitch cycles becomes apparent, as is shown in figure 1–23. Notice that the oscillations are not symmetrical in time but sawtoothlike, indicating that the cold phase built up steadily to a maximum level, then crossed a threshold and rapidly collapsed. Other studies that used different chemical techniques on coral skeletons have shown that the average annual surface sea temperatures in the Caribbean adjacent to Central America dropped 5 degrees centigrade 20,000 years ago, at the height of the last glaciation. Studies of pollen records on land (see chapter 5) show that similar changes in temperature were taking place on land.

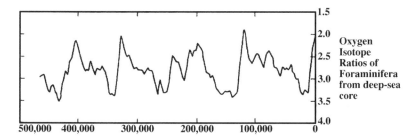

1–23. A diagram of the sawtooth pattern of sea temperatures during the Ice Age. The frequency of the pattern strongly coincides with predicted global heat fluctuations owing to Milankovitch cycles and to rises and falls of sea level. Lower isotope ratios are colder. Modified from J. D. Hays, John Imbrie, N. J. Shackleton, "Variations in the Earth's Orbit: Pacemaker of the Ice Ages," *Science* 194, no. 4270 (1976): 1130, fig. 9.

The final stages in the transformation of the Isthmus of Central America from a complex and extensive archipelago into a more simple isthmus were accompanied, then, by the onset of extensive glaciation in the northern hemisphere with major oscillations of sea level. Reconstructing the final moment of closure is no easy task.

One method is to detect in the fossil record the first evidence of one of the differences that now contrast the Pacific with the Caribbean. A good example is the strong seasonal temperature change in Pacific water in some locations. Trade winds drive surface water away from the isthmus from December to May, causing cold bottom water to rise and take its place—a process called *upwelling*. Animals such as mollusks secrete a layer of shell every month, and so the ratio of oxygen isotopes in each shell layer varies from the warmer wet season to the colder, dry upwelling season. Modern shells from the Pacific show this cycle clearly, whereas the same mollusks in the Caribbean, where there is no upwelling, show no such variation (fig. 1–24). Fossil shells that are 1.8–1.9 million years old (fig. 1–24) show the same contrast as modern shells, strongly suggesting that the isthmus was already formed 2 million years ago. In sediments that are older than 3 million years, however, the curves for the Pacific and the Caribbean are much more similar (fig. 1–24), indicating that there was less seasonality on the Pacific side. This suggests that the isthmus was not yet closed so that the wind-driven Pacific surface water could be replaced by warm Caribbean surface water. Chapter 2 describes in more detail the remarkable series of oceanographic and biological differences that have evolved between the two oceans in these past 3 million years.

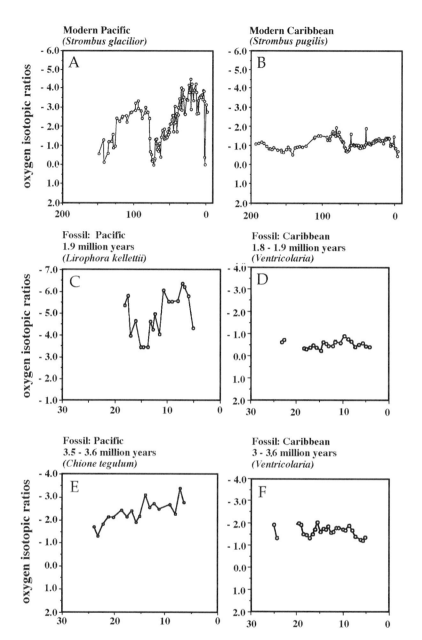

1–24. Diagram of the patterns of oxygen isotope measurements for monthly growth of mollusk shells (two-year cycle). The ratios correspond to temperatures of the seawater. (A, B) Living shells from the Pacific and Caribbean; (C, D) fossil shells 1.8–1.9 million years old from the Pacific and Caribbean; (E, F) fossil shells 3 million years or older from the Pacific and the Caribbean. Modified from Jane Teranes, "The Oxygen Isotope Record of Seasonality in Neogene Bivalves from the Central American Isthmus" in *Evolution and Environment in Tropical America,* J. B. C. Jackson, A. F. Budd, and A. G. Coates, eds. (University of Chicago Press, Chicago, 1996).

CHAPTER TWO

The Ocean Divided

JEREMY B. C. JACKSON and LUIS D'CROZ

The formation of the Central American isthmus was the pivotal event in the past 10 million years of earth history. Separation of the two oceans stopped the once strong westward flow of water from the Atlantic into the Pacific and gave birth to the Gulf Stream. The resulting strong poleward transport of tropical Atlantic water warmed the North Atlantic and increased precipitation, thereby providing moisture that intensified northern hemisphere glaciation. This triggering of the Ice Age by a distant change in tropical oceanography is one of the most dramatic examples of the sensitivity of global climate to geographical changes in the relation of continents and oceans.

The new barrier also changed the climate and oceanography on both sides of Central America and divided the once continuous tropical American ocean into two ecologically different realms. The eastern Pacific is highly variable and seasonal, with rich *pelagic* (open-sea) fisheries but poorly developed coral reefs, in contrast to the Caribbean, which has near-shore fisheries concentrated largely around extensive coral reefs. All of these systems are now profoundly threatened by man.

Central America is a maritime land, every nation but Belize and El Salvador being bordered by both oceans, and the ratio of coastline to land is the highest in the continental Americas. The differences in the two oceans are therefore of fundamental economic, social, and political importance and have been so for thousands of years.

Because of its unique history, Central America is an ideal model for understanding how geographic isolation or connection and environmental change affect life on earth. The dramatic consequences of the connecting of North and South America are well known (see chapter 4). In contrast, the effects of the separation of the Caribbean and eastern Pacific have received much less attention. Fossil species of marine invertebrates from Florida to California are broadly similar until about 5 million years ago but vary afterward. Likewise, chemical analyses of deep-ocean cores from the Caribbean and eastern Pacific demonstrate significant divergence in the temperature and salinity of the sea surface about the time of the final separation of the oceans. Until recently, however, researchers have lacked a precise geological framework in which environmental and biological events might be put in rigorous historical perspective.

In this chapter we compare similarities and differences in marine environments and marine life of the two tropical American oceans that border Central America and how they came to be. In addition to being of intrinsic scientific interest, the results help to explain differences in the vulnerability of the two oceans to human disturbance.

Two Different Oceans

The coastal geography of the two sides of Central America reflects their dissimilar geological origins (see chapter 1). The Caribbean coast is broadly sinuous and forms the western boundary of a semi-isolated sea; maximum distances between Central America and the Greater Antilles are only a few hundred to one thousand kilometers (fig. 2–1). The shelf is narrow along Panama and Costa Rica but widens broadly to more than 200 kilometers off the Nicaraguan and Yucatán peninsulas, whose offshore banks extend with only narrow breaks all the way to Cuba and Jamaica. In contrast, the eastern Pacific coast is comparatively straight and wide open to the Pacific, with the nearest substantial land being located more than 10,000 kilometers to the west. The continental shelf is narrow and closely bounded by the Middle America Trench, which brings deep-ocean conditions within 50 kilometers of the land everywhere except the Bay of Panama (see fig. 2–1).

Central America lies in the path of the trade winds, which blow strongly from the northeast to southwest all year. The seasonal climate of the region is determined by the interaction of the trade winds with a large, low-pressure air mass called the Intertropical Convergence Zone. This convergence zone moves northward to sit over Central America from approximately May to December each year (fig. 2–2), interrupting the flow of the trade winds and bringing variably intense rainfall,

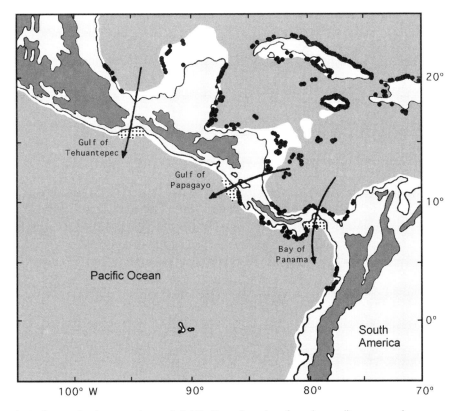

2-1. Geography, topography, and distribution of coral reefs and upwelling zones along the coasts of Central America. The Caribbean coast is part of the extensive western Caribbean system of coral reefs (black dots), mangroves, and sea grasses that blankets the coastline throughout most of the region. Arrows indicate passes in the mountains where trade winds blow across the isthmus and cause upwelling, shown by the stipple pattern. White areas are continental shelf between 0 and 200 meters depth; dark gray areas are land above 2000 meters.

which is everywhere greater along the Caribbean slope of the continental divide. In contrast, during the dry season from December to May (fig. 2–3) the Intertropical Convergence Zone lies well south of Central America, so the trade winds blow across the isthmus and precipitation is much less.

Caribbean coastal waters are warm and salty, with little annual variation in spite of very heavy rainfall most of the year. This constancy is due to the strong, westward flowing Caribbean Current, which bathes the southern Caribbean throughout the year (fig. 2–4). The tidal range is less than 1 meter and generally depends more on local weather than on any regular astronomical cycle. Weak upwelling of nutrient-rich waters

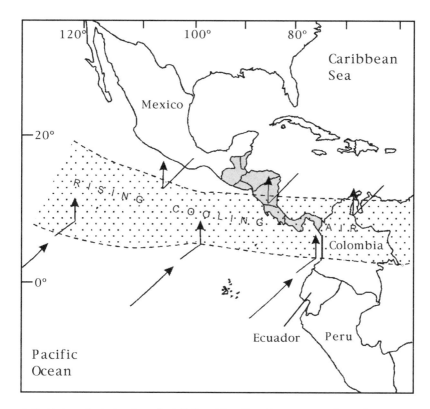

2–2. Meteorological pattern that distinguishes the rainy season in Central America. From about May to December the Intertropical Convergence Zone lies over Central America as shown. In this zone, air from the northeast and southwest trade winds meet and rise. The result is cooling of the air mass, condensation, and widespread rain.

(fig. 2–5) occurs off parts of Venezuela and Colombia, but not along the Caribbean coast of Central America, so levels of nutrients there are very low. Hurricanes occur with increasing frequency, intensity, and devastation northward from Costa Rica to Yucatán. Any particular storm, however, usually affects only a small part of the coast.

Seasonal climatic and oceanographic fluctuations are much stronger on the Pacific coast than on the Caribbean. This difference is pronounced in areas of wind-induced coastal upwelling adjacent to the Bay of Panama and the Gulfs of Papagayo and Tehuantepec, where the land is low and strong trade winds blowing across these lowland saddles drive the Pacific surface water out to sea (see fig. 2–1). This water is replenished by upwelling of underlying colder, denser water that is very rich in nutrients (see fig. 2–5). Surface seawater temperatures during upwelling may plummet more than 10 degrees in a few days to as low as

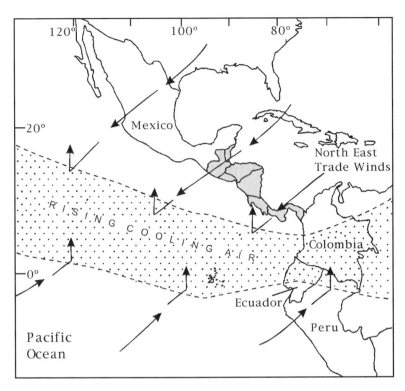

2–3. Meteorological pattern that distinguishes the dry season in Central America. From about January to April the Intertropical Convergence Zone migrates to the south, as the earth's axis changes orientation relative to the sun. The northeast trade winds then blow steadily across the Central American isthmus, bringing stable, drier, and less humid conditions.

15 degrees centigrade (pl. 1). Oceanic upwelling also occurs off the Pacific coast of Costa Rica owing to the divergence of the North Equatorial Counter Current. Pacific coastal salinities drop during the rainy season, especially in the Bay of Panama. The daily tidal range is large, reaching 6 meters or more on a regular semidiurnal and lunar schedule.

The pattern of ocean currents affecting the Pacific coast is also more variable than that in the Caribbean (see fig. 2–4). The main source of surface water is the Equatorial Counter Current, which flows eastward just north of the equator and then curves back northwestward from Panama as the Costa Rica Current. In addition, the Equatorial Counter Current is bounded by the North and South Equatorial currents flowing to the west, and these are fed by the California and Peru currents, which flow toward the equator along the coasts of North and South America, respectively. The California and Peru currents occasionally impinge on Central America as well.

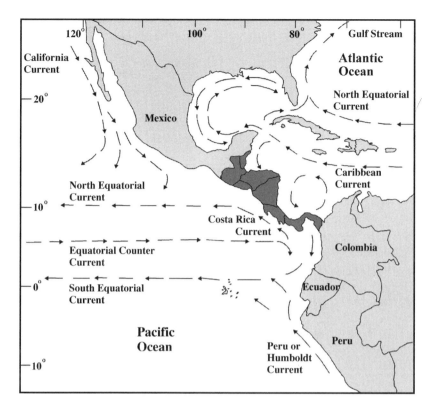

2–4. Names and flow patterns of the major ocean current systems that impinge on Central America.

This complex system is vulnerable to periodic disturbances in water movement, atmospheric pressure, and sea surface temperatures over the tropical Pacific. The resulting El Niño events occur every three to eight years and vary greatly in intensity. These events involve shifts in the relative position and strength of the various eastern Pacific currents and air masses which cause dramatic changes in rainfall, sea temperatures, upwelling, and biological productivity as well as changes in global climate from the tropics to the temperate zones. In Central America, these effects are greatest along the Pacific coast, where the dry season is extended, sea surface temperatures rise, and coastal upwelling is reduced.

These differences in oceanography between the comparatively stable Caribbean and the more variable eastern Pacific underlie equally striking biological variations. But to explain why, we must first consider the pathways of biological production in the sea and how organisms can alter the environments in which they live.

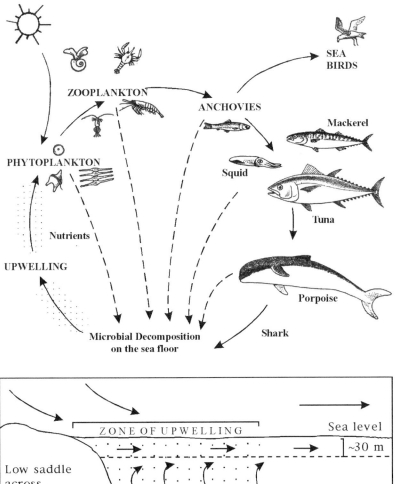

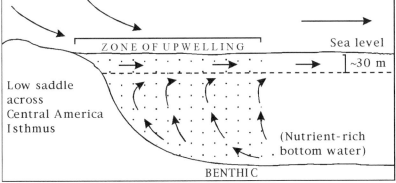

2–5. Upwelling and the pelagic food chain in the tropical eastern Pacific. Upwelling occurs opposite low saddles in the Central American volcanic mountain chain, where wind blows the surface water away from the coast. This water is then replaced by cool bottom water whose rich nutrients allow phytoplankton to multiply in abundance. The pelagic food chain, schematically shown here, is based on the abundant phytoplankton.

Production, Construction, and Environmental Variation

Biological production is the growth and multiplication of life, whereas biological construction is the assembly of biological materials, for example, the conversion of animal skeletons and wood into durable structures such as coral reefs and forests, respectively. Rates of biological production and construction respond to environmental changes in diverse ways.

Biological Production

Primary producers are organisms that do not require organic materials as a source of energy; all other life depends on them. Almost all *primary production* occurs by photosynthesis, whereby plants use chlorophyll and energy from the sun to convert carbon dioxide and water into carbohydrates and oxygen. Photosynthesis occurs only near the sea surface, where penetration of sunlight provides sufficient energy. In clear coastal waters sunlight reaches the sea floor down to 20 or 30 meters depth. In highly turbid waters the limit may drop to only a few meters, whereas in exceptionally clear ocean waters, photosynthesis may occur as deep as 100 meters. Thus, photosynthesis occurs exclusively in the water in the pelagic realm and reaches the sea floor only near land.

Almost all pelagic primary producers are extremely small, unicellular "plants" that drift passively in the current and are referred to collectively as *phytoplankton*. The most important groups are diatoms, dinoflagellates, coccolithophores, blue-green algae, and bacteria; the exception is the abundant, large, floating alga *Sargassum*, which gives the Sargasso Sea in the North Atlantic its name. In contrast, *benthic* primary producers live on the bottom and include typically much larger algae (collectively called seaweeds), sea grasses, and mangroves (an association of shrubs or trees which grow in saline soils in tidal fringes of tropical coasts). The most important exceptions are microscopic single-celled dinoflagellate algae that live symbiotically within the tissues of such plant-shaped animals as reef corals, sea fans, and sea anemones.

The ecology of pelagic and benthic primary producers is dramatically different. Phytoplankton abundance and production depend primarily on currents and upwelling, which supply essential nutrients like nitrogen and phosphorous. Growth and reproduction by simple cell division is rapid, and lifetimes are measured in days or weeks. Populations can therefore double within a day when nutrients increase and may fluctuate wildly in response to changing environmental conditions. Phytoplankton are eaten by *zooplankton*, which also drift with the currents. The most important of these primary consumers are microscopic copepod crustaceans, developing larvae of myriad invertebrates and fishes, and much larger jellyfish and salps (transparent, free-swimming tunicates or sea squirts). Zooplankton are eaten by small predators like anchovies, which in turn are eaten by larger and larger actively swimming predators, including tuna, squid, porpoises, and sharks (see fig. 2–5). Phytoplankton abundance is strongly affected by all of these consumers, but, in general, rates of production probably depend more on changes in nutrients, which means that the system is controlled from the bottom upward.

In contrast, the ecology of benthic primary producers depends more on biological interactions like competition for space, grazing, and disease than on currents and nutrients. Growth and reproduction are slower, and lifetimes are measured in months to centuries. Thus, populations cannot increase as rapidly to take advantage of newly favorable conditions, and numbers fluctuate within much narrower limits except when devastated by rare disasters such as hurricanes or epidemics. Benthic food chains are also more complex than pelagic food chains because there are more ways to feed. For example, such primary consumers as sea urchins, snails, parrot fish, and turtles feed on benthic primary producers like sea grasses and seaweeds, whereas sponges and oysters are suspension feeders that eat microscopic phytoplankton and zooplankton. These primary consumers are in turn eaten by a great variety of invertebrate predators, including worms, snails, and starfish as well as fish, sharks, and stingrays. In addition, a great variety of clams, snails, and worms feed on the detrital remains of dead organisms that accumulate on the bottom. In spite of this diversity, however, the outbreak or disappearance of a single top predator can totally alter the community. Thus, benthic communities appear to be controlled from the top downward except perhaps in strongly upwelling regions.

Biological Construction

Some phytoplankton and zooplankton produce minute skeletons of silica and calcium carbonate that accumulate as fine-grained sediments on the deep sea floor but are obscured in coastal environments by runoff of vastly greater amounts of sediment from the land. In contrast, many benthic plants and animals produce massive structures of wood, rhizomes (underground stems that anchor sea grasses), and limestone skeletons, which are the building blocks of the enormous biological constructions called mangrove forests, sea grass beds, and coral reefs.

Even sandy beaches and lagoonal sediments can be composed almost entirely of the fragmented calcareous skeletons of benthic animals and plants.

Coral reefs, sea grass beds, and mangroves occur together throughout most of the Caribbean (fig. 2−6), just as they do in the Indian and West Pacific oceans. In contrast, in the eastern Pacific, coral reefs and mangroves are effectively separated ecosystems and sea grasses are absent. In the absence of reefs, however, mangroves and sea grasses are restricted to bays and estuaries, where they are protected from the full force of the sea. When reefs are present, mangroves and sea grasses may occur behind them anywhere along the coast. The dense root systems and rhizome mats of mangroves and sea grasses reduce water circulation, increase sedimentation, and stabilize the bottom. In this way they act as sediment traps that protect coral reefs on wet coasts like Central America, where extensive runoff from the land might otherwise suffocate coral reefs in mud. Where they occur together, then, coral reefs, sea grasses, and mangroves are strongly interdependent: if one is disturbed, the effects are felt throughout the coastal ecosystem.

Where conditions are favorable, coral reefs, sea grasses, and mangroves dramatically change the physical structure and appearance of the coast (figs. 2−6, 2−7). The scale of their biological construction is enormous. Coral reefs are the largest and most durable construction projects on earth. The Panama Canal is still the largest human construction project, but it is paltry in comparison with the size and extent of coral reefs along the Caribbean coast of Panama, and these in turn are tiny compared to the Belize Barrier Reef, the largest reef tract in the Caribbean. Moreover, living reefs, including the Great Barrier Reef in Australia, have grown to their vast size in less than 7000 years, about the same time that humans have been building pyramids and cities. Investigators know this because sea level rose 85 meters between 12,500 and 7000 years ago, after the last glacial advance, including two 1000-year pulses during which it rose faster than 2 meters per century or 20 meters per thousand years! This is faster than coral reefs can grow, so that almost all reefs were drowned, and new reefs started to develop only when the rise in sea level slowed down.

The impressive size of coral reefs, sea grass beds, and mangroves gives a false impression of stability that belies an underlying turmoil of growth, death, and destruction of the organisms that construct these communities. For example, reef growth is determined by how much production, accumulation, and cementation of skeletons, especially of corals and coralline algae (a very important group that make skeletons

2–6. Caribbean coastal ecosystem along the north coast of Panama. Extensive fringing coral reefs are marked by the line of breaking waves to the right. Behind the reefs is a broad lagoon floored by turtle grass, inland of which is a broad zone of mangroves. Photograph by Carl Hansen.

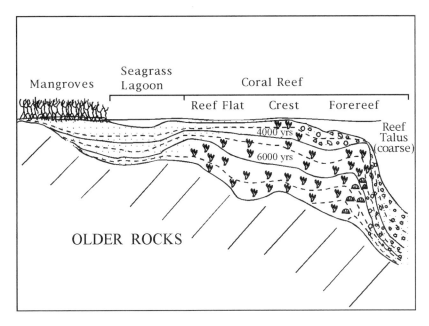

2–7. Idealized cross section of the internal structure of a Caribbean mangrove / sea grass / reef coastal ecosystem. Dashed lines show the profile of the reef at various stages in its history; the profiles for 4000 and 6000 years ago are picked out in solid black lines.

2–8. Processes of reef destruction: (A) accumulation of sediments killed the central portion of this lettuce coral (*Agaricia sp.*); (B) the black-spined sea urchin *Diadema antillarum* has scraped away most of the large brain coral at the center; (C) the tiny (5-centimeter-long) three-spot damselfish *Stegaster planifrons* (arrow) is killing the coral in the background by biting away tissue to expose the skeleton beneath (white spots). This allows algae to grow on the skeleton. These algal gardens provide food for the fish, who vigorously protects them from intruders; (D) the large midnight parrotfish *Scarus rubrovidaceus* bites away large pieces of coral with its strong beak (white spots on the coral below). Photographs by Carl Hansen.

in the manner of corals), exceed destruction due to storms (pl. 2D), grazing, and excavation by animals and plants (*bioerosion*) that can penetrate the reef limestone chemically or physically (fig. 2–8). Similarly, expansion of mangrove forests and sea grass beds depends on the ability of these plants to trap, stabilize, and accumulate sediment to grow on. The crucial point is that the distribution and extent of these communities are sensitive to even small changes in rates of construction or destruction. For example, most eastern Pacific coral reefs virtually disappeared within a decade after the strong El Niño of 1983 because of coral death and intense bioerosion. There is evidence now of recovery in many areas.

Most of the primary production of sea grass beds and mangroves is due to photosynthesis by the bioconstructing organisms themselves or, in the case of corals, by their symbionts, rather than by pelagic phytoplankton. Much of this production is directly invested in the physical structure of the habitat in the form of wood, rhizomes, or limestone skeletons that tend to be recycled within the coastal ecosystem rather than exported to other environments. Net productivity and biomass are greatest for mangroves and least for coral reefs. Productivity of phytoplankton is variable in mangroves but low over sea grass beds and coral reefs, except in cases of overenrichment by man. Low nutrient content appears to be critical for extensive reef growth, and this sensitivity to nutrients is a major determinant of the differences in ecological communities on the opposite coasts of Central America.

Production and Construction in the Two Oceans

How, then, do all of these differences in biological production and construction relate to the physical oceanographic variations between the two coasts of Central America? The two critical factors are the amount of nutrients available and the physical stability of the environment. In general, high pelagic nutrient concentrations result in dense populations of rapidly multiplying phytoplankton that decrease light penetration and rob nutrients from slower-growing benthic primary producers. Similarly, high nutrients near the bottom greatly favor seaweeds and animals without massive skeletons over corals, sea grasses, and mangroves because they grow much faster than the latter. Rapidly fluctuating environments also tend to tip the balance in the same direction as high nutrients.

Pelagic Systems

Pelagic primary production during seasonal upwelling in the Bay of Panama and the Gulfs of Papagayo and Tehuantepec is the highest in Central America and supports extremely rich pelagic fisheries that may produce half a million metric tons per year. Anchovies account for most of this biomass (pl. 3A, B). These small fish feed directly on the plankton. Their abundance varies from as little as a few thousand tons annually when El Niño events suppress upwelling to as much as 200,000 tons when upwelling is very strong. Anchovies and other small fish in turn feed abundant tuna, dolphins, and seabirds like brown pelicans and boobies (see fig. 2–5, pl. 3A) that nest by the millions in the Bay of Panama during upwelling. When upwelling failed during the extreme El Niño of 1983, every baby pelican died. Comparatively little is known about pelagic fisheries along the Caribbean coast of Central America, largely because low nutrients and lack of upwelling result in such low abundance of pelagic fish that no one has bothered to study them in detail. Caribbean seabirds are also rare.

Benthic Systems

The sandy, muddy bottoms of estuaries, bays, and the continental shelves are inhabited by diverse populations of worms, clams, snails, shrimps, crabs, lobsters, sea urchins, starfish, and flatfish like flounders. These broad sediment plains are physically unstable and extensively burrowed by animals, so that bioconstruction is inhibited. The composition of these benthic communities is strongly influenced by levels of production in the waters above. In the eastern Pacific, abundance and spawning of scallops and other fast-growing shellfish that feed directly on plankton and detritus increase dramatically during most strong upwelling years. Mollusks in general are extremely abundant, and dredge hauls in the eastern Pacific are almost always full of shells, which is rarely the case in the Caribbean.

The extent of the two Central American coasts dominated by corals, sea grasses, and mangroves is exactly the opposite of that of productive pelagic and benthic fisheries. Large coral reef tracts, sea grass meadows, and mangrove forests border about half of the Caribbean coast (see fig. 2–1). The chief exceptions are the Mosquito Gulf in Panama, most of Costa Rica, and the Mosquitia of Honduras and Nicaragua, where sediment discharge from large rivers inhibits near-shore development of reefs. In contrast, eastern Pacific reefs are small and widely scattered, sea grasses are absent, and mangroves occupy less than one-third of the area they cover in the Caribbean.

Wood and tough skeletons provide structural support and protection against most predators and grazers. For example, only animals like parrot fishes and sea urchins that are armed with strong beaks can scrape through the skeleton of corals and other stony species to graze on the soft tissues below (see fig. 2–8B–D), when other more palatable foods are not available. On the other hand, making a skeleton takes time and costs energy, so corals grow slower and reproduce less than most seaweeds or sponges, which do not invest so heavily in skeletons. Moreover, skeletons are not the only form of protection against predators. Many seaweeds, sponges, and other attached (*sessile*) animals produce toxins that make them unpalatable or poisonous.

These trade-offs among skeletal support, protection, and growth rate, along with the differential effects of nutrients on phytoplankton and corals, help to explain why coral reefs are extensive on the Caribbean coast of Central America but not the eastern Pacific. Sessile animals that feed on phytoplankton in the water—for example, sponges, bryozoans (small, colonial, mostly calcified marine animals), and oysters—grow and reproduce extremely rapidly on the Pacific coast, especially in zones of upwelling. Moreover, because abundant nutrients favor sea-weeds as well as phytoplankton, both seaweeds and plankton-feeding animals tend to rapidly colonize and overgrow substrata suitable for corals, and reefs are small and ephemeral. Similar arguments may apply to sea grasses, which are absent from the tropical eastern Pacific.

In contrast, low nutrients on the Caribbean coast favor the coral-algal symbiosis because the symbiosis recycles scarce nutrients much more effectively than seaweeds or phytoplankton, and corals are also better protected against grazing parrot fish and sea urchins by their stony skeletons. The greater environmental stability of the Caribbean also favors longer-lived corals over shorter-lived seaweeds and animals. Nonupwelling regions of the Pacific coast like the Gulf of Chiriquí present an intermediate situation. Here, productivity and environmental fluctuations are less common than in upwelling regions but greater than in the Caribbean and, as might be expected from the above arguments, can support only reefs of intermediate development.

The Community Ecology of Coral Reefs, Sea Grasses, and Mangroves

Coral reef, sea grass meadows, and mangrove forests are built by relatively few species. In Central America, there are fewer than 100 species of corals, a dozen species of mangroves, and half a dozen sea grasses; even the Great Barrier Reef has only about 600 species of corals. Thus, the very existence of these habitats depends on the well-being of relatively few species. Moreover, Central American reefs, sea grass beds, and mangroves provide habitat and shelter for thousands of species of fishes, shrimps and other crustaceans, starfish, sea urchins, snails, clams, oysters, worms, bryozoans, sponges, algae, and myriad other groups (fig. 2–9B); just as canopy-forming trees provide habitat for the spectacular array of birds, insects, mosses, and herbs of a tropical forest.

Most of these associated species exhibit preferences for specific reef, sea grass bed, or mangrove habitats, where they may spend their entire lives. Others begin life in one habitat and migrate elsewhere when they are big enough to escape from predators and defend themselves or are better equipped to eat different foods. The migrators include commercially important shrimps and fishes like snappers, which explains why mangroves are so vital as nurseries to fisheries in other habitats, such as those along the Pacific coast, where sea grasses are ab-

2–9. Mangrove root systems: (A) large red mangrove roots (*Rhizophora mangle*) on the Caribbean coast of Panama tower above the scientist below. The maze of roots protects the coast from waves and traps sediment running off from the land; (B) abundant edible oysters growing underwater on a red mangrove root in the Chiriquí Lagoon, Bocas del Toro, on the western north coast of Panama. Photographs by Marcos Guerra.

sent and reefs few and small. Very few species except the larger vertebrates range freely among habitats throughout their lives.

Many species that live among corals, sea grasses, and mangroves have little positive or negative effect on their habitat. But others eat, clean, and protect the corals, sea grasses, or mangroves on which they depend and can therefore greatly affect the distribution and abundance of their hosts. The best studied examples of these *keystone species* are crabs that feed on mangrove seedlings (pl. 4C), turtles, manatees, and sea urchins that feed on sea grasses (fig. 2–10), and starfishes, snails, and fishes that feed on corals and algae (see fig. 2–8B–D).

Coral Reefs

Reef corals are cylindrical polyps similar to sea anemones that secrete a hard, calcareous skeleton at their base (pl. 5A). Most corals grow by repeated budding of polyps, which remain attached to form a colony (pl. 5B–D). Corals are animals, but all reef-forming species contain extremely abundant symbiotic algae within their tissues, and the same is

2–10. Historically abundant, large Caribbean grazing vertebrates: (A) the manatee *(Trichechus manatus)* in Nicaragua, a species approaching extinction in many locations. Photograph by Jim Reid, U.S. Department of the Interior, Sirenia project; (B) green turtles *(Chelonia mydas)* nesting at Tortuguero on the northeast coast of Costa Rica. Photograph by U.S. Department of the Interior. Turtle populations are now vastly reduced throughout the coastal waters of Central America.

true of sea fans and most of the other relatives of corals that inhabit reefs. Photosynthesis by the symbiotic algae during daylight supplies most of the energy for coral growth. Even under normal conditions, however, the polyps also feed as animals, using their tentacles to catch zooplankton, because the food that the algae provide is "junk food," insufficient in protein to meet all of the coral's needs.

The symbiosis between corals and their algae is extremely complex but is normally stable over a wide range of environmental conditions. Reef corals placed in the dark stop growing, release their algae, and eventually die. Similar "coral bleaching" occurs in natural reef populations when the coral expels its algae (or perhaps the algae abandon ship) when the water is too hot or too fresh, among other factors (pl. 6). At present, there is great controversy about whether the frequency of coral bleaching is increasing and is therefore a harbinger of global warming. Whatever its cause, however, bleaching is definitely a sign of stress. High concentrations of nutrients in the water are also unfavorable for corals because they disrupt the regulatory balance between the algae and their coral hosts and increase photosynthesis by phytoplankton.

The plantlike shape of corals is determined by patterns of budding of the polyps (see pl. 5B–D), which vary among species and, to a lesser extent, in response to changes in light and water movements. The most common forms are branching, massive, and foliaceous. In the Caribbean, branching corals typically dominate from the reef crest down to 5 or 10 meters (see pl. 2A,B), at which point they are replaced by mostly massive, dome-shaped species that in turn give way to flat, platelike forms that extend downward to 50 meters or more. Zonation is similar in the eastern Pacific but diversity within any zone is typically much less. Branching corals grow up to ten times faster than brain corals and other skeletally massive forms, but there is also great variation among branching species. The champions are the staghorn and elkhorn corals in the genus *Acropora*, which grow from 10 to nearly 30 centimeters per year. These are the competitively dominant species on wave-swept Caribbean shores (see pl. 2A, B) but are absent from the eastern Pacific. In contrast, smaller branching finger corals like *Pocillopora* and *Porites* grow only 2 to 3 centimeters per year. Nevertheless, *Pocillopora* is extremely abundant on eastern Pacific reefs, where it commonly forms dense, single-species stands in shallow water (see pl. 2C); branching *Porites* commonly do the same on more sheltered Caribbean reefs.

Catastrophic declines of reef corals have occurred around Jamaica and at many other sites throughout the Caribbean during the past ten to twenty years. Extreme overfishing in Jamaica had long ago resulted in a severe decrease in parrot fishes and other grazing fishes that consume seaweeds. These fish were replaced ecologically, however, by extremely abundant, long-spined sea urchins and other small invertebrate grazers that kept the seaweeds down (fig. 2–11A). Then in 1983 an unidentified plague swept through the Caribbean, killing more than 95 percent of the long-spined sea urchins (fig. 2–11B). This left seaweeds free to overgrow the corals. During the following decade, the coverage of live corals on Jamaican reefs plummeted to only a few percent of that before the plague. Now the same thing is slowly happening all along the Caribbean coast of Central America wherever the deadly combination of overfishing, increased nutrients due to deforestation and fertilizers, and sea urchin death is tipping the balance of power to seaweeds over corals. These effects of overfishing and sea urchin mortality are dramatic examples of the importance of keystone species on coral reefs.

Sea Grass Beds

Caribbean sea grasses are most abundant from the intertidal zone down to about 10 meters (see pl. 4A, D), although they may reach 30 meters in areas of exceptionally clear waters; different species are abundant at different depths. The most common species are the broad-leaved turtle grass and slender-leaved manatee grass, common names that evoke the abundance of animals that have practically disappeared from these ecosystems. Sea grasses form extensive carpets of green leaves that grow upward from the sediment and become more ragged and encrusted toward their tips. Turtle grass leaves may grow more than half a meter long before they break off, but more commonly they are cropped down by grazers to less than half that length. Leaf growth may be as fast as 1 centimeter per day, and gross primary production rivals that of a field of corn. Beneath the surface, turtlegrass builds an extraordinarily dense, nearly impenetrable mat of roots, stems, and rhizomes that stabilizes the bottom and helps to protect burrowing animals from being excavated by stingrays and other benthic predators.

Numerous animals feed directly on sea grass leaves and considerably more feed on sea grass detritus. The most important leaf eaters today are sea urchins, which can destroy entire sea grass beds, and small herbivorous fishes. Only a century ago, however, the most important grazers on sea grasses were green turtles, manatees (see fig. 2–10), and much larger fishes. Columbus and other early voyagers were amazed by the abundance of turtles that littered the vast turtle grass beds of the Mosquitia Bank and elsewhere like so many millions of stones. In ad-

2–11. Black-spined Caribbean sea urchin *Diadema antillarum* : (A) grazing horde on a reef pavement before the massive die-off in 1983; (B) a specimen dying from the unknown pathogen that killed more than 90 percent of the species throughout the entire Caribbean. The urchin's spines on top have fallen off, exposing bare flesh beneath. Photographs by Haris Lessios.

dition, the giant queen conch processes large amounts of turtle grass and its detritus in order to feed upon the associated microorganisms on the blades. Early descriptions of conchs are not so dramatic as those of turtles, but their shell middens (piles of shells from which the meat has been extracted) are so enormous that they tower above the surrounding landscape at some once-rich conch fishing grounds.

Conchs are rare now because of overfishing, but when placed in dense experimental populations they damage or destroy the turtle grass while trying to feed; and the abundance of conchs in ancient middens suggests that they may have had a similar natural effect before overfishing. Feeding by green turtles is potentially even more destructive, although by processing vast amounts of sea grass leaves through their guts turtles must have greatly increased the rate at which sea grass entered the detrital food chain. Indeed, the extensive migrations of green turtles probably evolved because dense local populations of turtles were unsustainable, just as wildebeest must migrate on the East African grass plains. Only 150 years ago, a large part of the productivity of sea grasses was harvested as turtle, manatee, fish, and conch meat by people; but today, that productivity has been lost to small, mostly unpalatable invertebrates. This is another example of the importance of keystone species in communities dominated by bioconstructional species.

Mangroves

Mangrove forests mark the boundary between marine and terrestrial environments in tropical zones in the same way that salt marshes do in the temperate zone (see fig. 2–9A; pl. 4A, B). The most important factors for mangrove development and productivity are low wave energy, protected shorelines, abundant freshwater and nutrients, mixed salinities, and the deposition and accumulation of fine organic mud. All of these conditions occur along both coasts of Central America, although the absence of large reef tracts in the eastern Pacific restricts mangrove development to bays and estuaries, where they are protected from the open sea. In contrast, the much larger tidal amplitude on the Pacific coast allows mangroves to extend farther inland than in the Caribbean.

Zonation of different mangrove species is due to the interaction of all these environmental factors as well as to grazing on mangrove seedlings and catastrophic storms. Consumption of seedlings is almost entirely by crabs (see pl. 4C) and snails, although large vertebrates may also have been important previously. Rates of grazing vary greatly among different mangrove habitats, forests, and regions. For example, in a recent study, crabs in Panama and snails in Florida consumed about 10 percent of experimentally tethered seedlings every four days. In contrast, consumption rates in Malaysia and Australia were 25 percent over a four-day period, making it quite remarkable that any seedlings ever survive! Consumption rates may also be high on the Pacific coast of Central America, where similar crabs and snails are commonly more abundant than in the Caribbean. Other experiments in Australia have also shown that mangrove species can colonize areas in which they are normally absent when herbivores are excluded by cages. Thus, the zonation of mangroves is controlled more by these small animals than by the direct effects of the physical environment, which is yet another example of the importance of keystone species.

Export of the biological production of mangroves from the forests toward the coast occurs mostly as detritus derived from mangrove leaves, and secondarily as the bodies of abundant fishes and shrimp that live as juveniles in the mangroves before migrating elsewhere. Leaf production may be as high as 10 to 15 metric tons per hectare per year but is not immediately available as food for animals until the leaves are converted into detritus. This conversion involves the leaching of lignin and tannic acids, biological transformation through the metabolism of bacteria and fungi, and physical manipulation of the leaf fragments by myriad worms and small crustaceans. After several weeks only a small fraction of the original leaf remains, but the concentration of nutritious compounds is increased more than tenfold. Mangrove leaf detritus is the most important food for many coastal species, which is why the volume of fisheries in a region increases with the amount of mangroves present.

The Fossil Record of the Oceans

The marine fossil record of Central America is extremely rich but is more complete for the Caribbean coast than for the Pacific (fig. 2–12). Most of the Pacific record has been subducted beneath the tectonically active Pacific margin. Structural deformation of marine sediments is also greater on the Pacific side, with the result that preservation and stratigraphic ordering of Caribbean fossils are generally much better than those of the Pacific. For these reasons we will be largely discussing events on the Caribbean side of Central America.

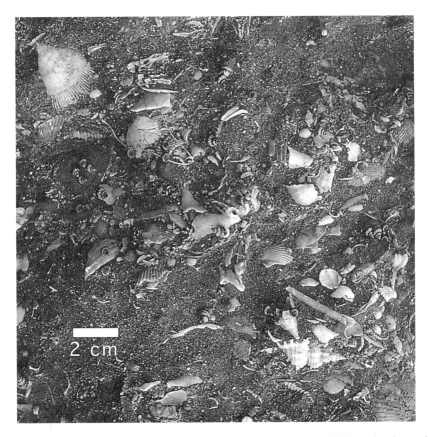

2–12. Abundant fossil snails, clams, tooth shells, and corals (*upper left*) on the shore of Cayo Agua, Chiriquí Lagoon, Panama. The fossils are from the Pliocene epoch and are about 3.5 million years old. The small black-and-white snails are living periwinkles. Photograph by Marcos Guerra.

Changes in Diversity through Time

The diversity of Caribbean fossils is extremely high, as shown for mollusks from Costa Rica and western Panama (fig. 2–13). The curves are for hundreds of collections of fossils sorted into four time periods over the past 12 million years. Each curve plots the cumulative numbers of different genera against the numbers of specimens collected. These are the largest such collections ever made from Central America, but even 75,000 specimens are not enough to ensure that the full complement of common mollusks have been identified. Paleontologists know this because additional samples continue to unearth new kinds of mollusks for each time period. Only when new collections do not significantly

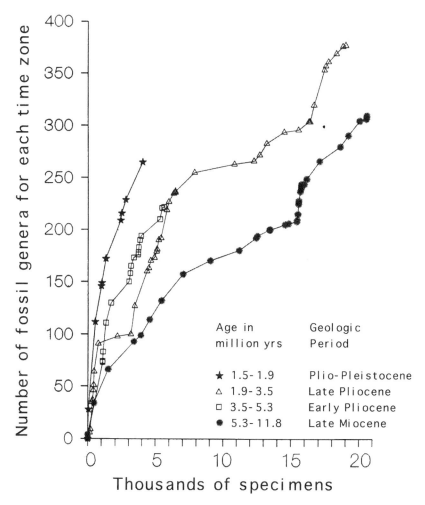

2–13. Diversity of fossil mollusks from four different ages that were collected along the Caribbean coast of Costa Rica and Panama. The numbers of different kinds of mollusks from each age are still increasing sharply as new specimens are obtained.

add new genera and the slope of the curves levels off will the true diversity be known. The oldest fauna is apparently the least diverse because the rate of increase is much less, whereas diversity for the three younger faunas appears roughly similar.

The first and last stratigraphic occurrences were determined for 1237 genera of mollusks identified as living during the past 12 million years. The purpose was to calculate rates of origination and extinction of molluscan genera through time and to determine whether these were constant processes or whether some times witnessed especially high rates of change. There was a striking pulse of origination and extinction about 2 million years ago, when nearly half of the earlier forms became extinct but were replaced by new ones within a few hundred thousand years (fig. 2–14). Before and after the turnover, however, rates of evolution were much less. Rapid turnover of mollusks also occurred at about the same time in Florida, where it was accompanied by a dramatic turnover of seabirds and marine mammals.

Species of Caribbean reef-building corals also experienced a strong evolutionary pulse about 2 to 3 million years ago, when rates of extinction increased more than tenfold. Turnover of corals, however, began earlier and was slower than for the mollusks. Only one-third of more than 100 coral species alive before the turnover survived, and extinction and speciation dropped sharply afterward so that the modern Caribbean coral reef fauna is about 2 million years old. The period of coral turnover also saw the explosive increase in abundance of branching corals in the genus *Acropora,* which grows more than twice as fast as any other corals and now dominates most living shallow water reefs around the world (see pl. 2A, B), except in the eastern Pacific.

Similarly comprehensive data for mollusks and reef corals are unavailable from the Pacific coast, so it is not possible to compare events

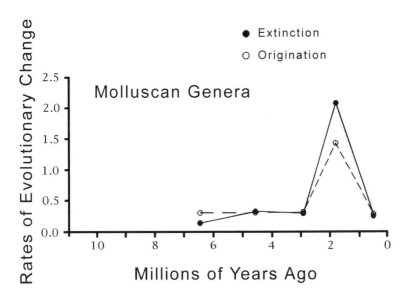

2–14. Curves showing the rate at which genera of mollusks have appeared and gone extinct during the past 10 million years. A sharp pulse of extinction coincides with a peak of origination two million years ago. A similar curve for coral species shows a peak of origination spread over 1 to 2 million years.

on the two sides of Central America. There are, however, excellent collections from both coasts of a highly diverse and well-studied family of snails called strombinids. These are mostly small, live on sand or mud, and have an exceptionally rich fossil record of more than 100 tropical American species. Strong turnover of strombinid species occurred on both coasts but in different ways. Initially there were many more species in the Caribbean than in the eastern Pacific, but the pattern was reversed about 2 million years ago owing to a pulse of extinction in the Caribbean and one of origination in the Pacific. The history of reef corals was also very different in the two oceans. Large *Acropora* replaced small *Pocillopora* finger corals as the dominant branching species in the Caribbean but never became established in the eastern Pacific, where *Pocillopora* dominates all other species (pl. 2C).

The very different histories of reef corals and strombinids on opposite sides of Central America may not be typical. Most transisthmian pairs of closely related sea urchins, shrimps, snails, and fishes are so similar morphologically that only taxonomists can tell them apart (fig. 2–15). Greater resolution is possible by fingerprinting the DNA of sister species on opposite sides of the isthmus, but even these molecular differences are relatively small. Thus, it appears that evolution due to geographic isolation alone is rather slow, especially compared to the pulse of origination and extinction affecting corals and mollusks 2 to 3 million years ago.

The most complete biological data for how the rising isthmus affected the timing of divergence are those for seven pairs of sister species of snapping shrimps, one of each pair living on the Pacific coast and the other on the Caribbean (fig. 2–15C). The extent of divergence of these species was measured by molecular differences in proteins and DNA and by the amount of aggression in experimental combats. Close correspondence of all three parameters in each species pair strongly suggests that they measure time since divergence in a clocklike manner (fig. 2–16).

Isolation of pairs of snapping shrimps did not occur at the same time. The four most similar pairs probably separated 3 million years ago, when the last marine connections were severed. If one assumes constant rates of divergence, then the most dissimilar pairs must have started to diverge more than 7 million years ago, when the oceans were still connected by shallow seaways. These results coincide closely with the ecology of the shrimps. The four most similar pairs occur in the midintertidal zone along the mainland, presumably one of the last habitats to be isolated by the isthmus. In contrast, the most divergent

2–15. Pairs of sister species from the eastern Pacific (*left*) and Caribbean (*right*) coasts of Panama: (A) pencil sea urchins: (*left*) *Eucidaris thouarsi*; (*right*) *Eucidaris tribuloides* . Photographs by Haris Lessios; (B) tube blennies: (*left*) *Acanthemblemaria castroi*; (*right*) *Acanthemblemaria rivasi*. Photographs by Ross Robertson; (C) snapping shrimps: (*left*) *Alpheus cylindricus* ; (*right*) *Alpheus cylindricus*. Photographs by Carl Hansen.

pairs require deeper water or more offshore habitats, which would have been isolated earlier.

Evolution and Environment

The progressive isolation of the two oceans by the emerging isthmus caused major changes in primary productivity in the two oceans that are correlated with origination and extinction. Three sorts of evidence suggest that primary productivity was high in both oceans until about 3 million years ago, when it increased in the eastern Pacific and de-

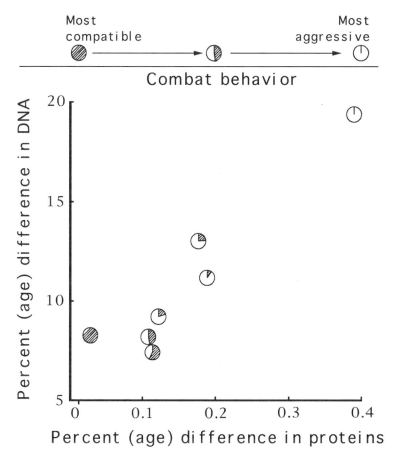

2–16. Genetic and behavioral divergence between species pairs of snapping shrimp whose ancestors were separated by the formation of the isthmus. The data points form a straight line, showing that sister pairs have the same level of DNA differences as they have protein differences; the pattern of combat behavior also goes from compatible to aggressive as the differences in DNA and protein get larger. The correlation between the three different kinds of measurements suggests that they measure time like a clock; the species pairs with the largest differences have been separated the longest. Modified from N. Knowlton, L. A. Weigt, L. A. Solórzano, D. K. Mills, E. Bermingham, "Divergence in Proteins, Mitochondrial DNA, and Reproductive Compatibility across the Isthmus of Panama," *Science* 260, no. 5114 (June 11, 1993): 1629–32.

clined dramatically in the Caribbean. First, fossil mollusks, seabirds, and marine mammals are exceptionally abundant from Florida to California prior to 3 million years ago and then decline in the Caribbean but not in the Pacific. Second, microfossils and sediments rich in phosphates (deposits that are characteristic of upwelling regions) disappear from the Caribbean at about the same time. Third, fluctuations in oxygen isotopes from within annual bands in fossil mollusk shells have been used to infer seasonality and productivity (as described in chapter 1). Some of these isotopic data suggest a gradual decline in Caribbean productivity and others a rapid decline, while eastern Pacific productivity increased.

Surface temperatures of both oceans must also have declined by several degrees during the time of faunal turnover because of the intensification of northern hemisphere glaciation. This conjecture is supported by isotopic analyses of the elements strontium and calcium and by the restriction of latitudinal ranges of tropical Atlantic species. Fossil and paleoenvironmental evidence are not yet sufficient to resolve whether changes in temperature or productivity were the chief agents of turnover: probably both were important.

The Effects of Humans

Less than 1000 years ago in western Panama, the Caribbean people living on the shores of the Chiriquí Lagoon derived most of their nutrition from the sea. Their garbage dumps are full of remains of turtles, fish, and shellfish, and remains of terrestrial animals and plant crops are comparatively rare. At about the same time, however, Pacific people living along the Gulf of Chiriquí consumed mostly terrestrial prey and plant crops like maize. These patterns are now completely reversed. Tourists in mainland Caribbean hotels innocently consume great quantities of what they think is fresh corvina, snappers, and groupers, which in reality are trucked over the mountains each day from the Pacific coast because Caribbean snappers and groupers are virtually gone. Scuba divers and snorklers in the same hotels have to be satisfied with beautiful but tiny reef fishes that dart about isolated remnant corals on reefs covered by piles of seaweed that have smothered once luxuriant corals. Offshore, on islands like Roatán, San Andrés, and Providencia, live corals are still abundant, but the fish are mostly gone and destruction is imminent.

The Pacific coastal communities of Central America are naturally more variable than those on the Caribbean, which probably explains why the effects of humans have been so much greater in the Caribbean. This is not to say that the Pacific coast has been unaffected. Much of the Pacific mangroves have been cut down for wood and to make way for human settlement or mariculture. The consequences of these losses are unknown, but it is not unlikely that the increased returns from shrimp farming, for example, are more than offset by decreases in fisheries offshore. It is difficult to understand how such mariculture projects can be implemented without attempting to demonstrate their projected costs and benefits to the economy as a whole.

The damaging effects of trawling are also unstudied in Central America, but extensive research in northern Australia has shown that the habitat structure is destroyed and that production shifts from desirable to undesirable species. Intensive trawling for shrimp has almost certainly transformed bottom communities in places like the Gulf of Nicoya and the Bay of Panama (see pl. 3C, D), where many formerly abundant species are absent or rare. For example, scientists have been unable to collect species of snails that were dredged commonly in the Bay of Panama during the first half of the century. Intensive trawling also occurs all along the Mosquitia Coast, but the effects are unknown.

In general, however, human effects on Caribbean marine communities are obvious and catastrophic. That the loss is so manifest is due to the communities' great bioconstruction, which still dominates Caribbean coastal processes even though it is in serious danger of collapse. Almost everything people do along the Caribbean coast is harmful to this bioconstruction, although adequate planning and management would reduce such damage. Uncontrolled deforestation (see pls. 15, 16) and increased land use for housing and agriculture without soil conservation measures increase sedimentation and nutrients that harm corals, increase bioerosion, and favor seaweeds and phytoplankton. Increased runoff of chemical fertilizers and sewage and overfishing do the same. As the habitat structure is destroyed, surviving fish depart, and the community is taken over by small invertebrates.

Bioconstructional communities are strongly interdependent, so that once destroyed, a great many things must happen for recovery to begin. The intricate nature of such interconnections was demonstrated after a major oil spill dumped nearly one-third as much oil as the Exxon Valdez spill in Alaska into a complex coastal lagoon at Bahía las Minas along the Caribbean coast of Panama (pl. 7). The oil quickly killed a band of mangroves along the shore, several sea grass beds, and most of the living coral on reefs that received heavy oil deposits (pl. 8A, B). But this was not the end of the story. Most of the oil that was not recovered was buried in the mangrove sediments. Erosion of these sediments increased as the dead mangroves and sea grasses rotted away, exposing the oil-soaked shore. Heavy rains at the beginning of each rainy season flushed the newly exposed oil and sediments out into the bay, causing new oil spills, injury, and death (pl. 8C). Sedimentation on reefs in-

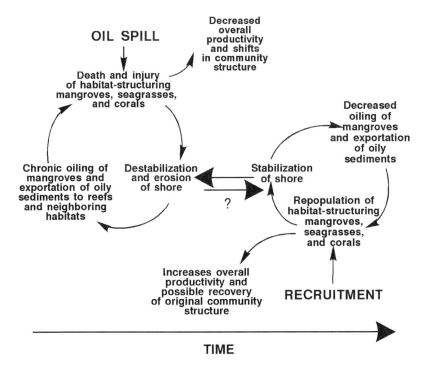

OIL SPILL

Decreased overall productivity and shifts in community structure

Death and injury of habitat-structuring mangroves, seagrasses, and corals

Chronic oiling of mangroves and exportation of oily sediments to reefs and neighboring habitats

Destabilization and erosion of shore

?

Stabilization of shore

Decreased oiling of mangroves and exportation of oily sediments

Repopulation of habitat-structuring mangroves, seagrasses, and corals

Increases overall productivity and possible recovery of original community structure

RECRUITMENT

TIME

2–17. Model of habitat damage and recovery after a major oil spill. Stabilization of the shoreline by mangroves and sea grasses is necessary for corals to become reestablished. Reprinted from B. D. Keller and J. B. C. Jackson, eds., "Long-term assessment of the oil spill at Bahía Las Minas, Panama," synthesis report, volume II: technical report. OCS Study MMS 93–0048. U.S. Department of the Interior, Minerals Management Service, Gulf of Mexico OCS Region, New Orleans, La., 1990, 1,017 pp.

creased fourfold over as many years, and there has been virtually no recovery of corals, sea urchins, or oysters at any of the most affected sites.

Although the causes are different, the consequences of this oil spill serve as a model for thinking about the long-term consequences of other, more pervasive human destruction (fig. 2–17). Bioconstruction depends on the stability of the shore, low levels of nutrients that suppress nonconstructional species like plankton and seaweeds, and abundant predators and grazers that consume potential competitors. Moreover, because adjacent systems interact, serious disturbance to one sets off a chain reaction of events harmful to the rest, so that the entire ecosystem degrades more and more. Even worse, all of the important bioconstructing organisms grow very slowly, as do the turtles, groupers, and sharks that used to dominate the food chain in these communities. Thus, the pace of recovery is extremely slow. Reversal of these events will require not only much more careful national and regional planning, but also the reevaluation of economic, social, and cultural priorities for almost everything people do along the coast! The alternative is the loss of the entire productivity of the Caribbean coast for human consumption and tourism and increasing erosion of the shore.

The one thing that human activity has apparently not yet seriously affected in Central American marine communities is the isolation of Caribbean and Pacific species. The Panama Canal is a freshwater canal, and plans for a sea-level canal have been widely discredited because of their cost and effects on ground water as well as other environmental risks. A far greater threat than a sea-level canal, however, is emerging from the indiscriminant transport of organisms in the seawater and sediments of the enormous ballast tanks of modern ships. Such unwitting importation has already resulted in invasions by exotic species worldwide, especially in estuaries and bays. The consequences can be highly destructive, as in the case of introduced parasites or predators that prey on commercially important shellfish and toxic phytoplankton that cause fish kills and red tides.

Similar introductions of transisthmian species could dramatically alter Central American coastal communities. Several species are of particular interest for their demonstrated ecological importance. Such predators on corals as puffer fishes and snails are larger and more voracious in the eastern Pacific. In addition, the crown-of-thorns starfish *Acanthaster planci* (fig. 2–18A) is absent from the Caribbean, although it readily consumes Caribbean corals when they are offered in experiments. The crown-of-thorns is famous for undergoing population explosions in the Indo-Pacific region, which produce hordes of starfish that consume almost all living corals before dying themselves of starvation. The reason for these starfish eruptions is strongly debated, and tens of millions of dollars have been spent in Australia alone trying to understand what sets them off. The basic question is whether they are natural events or are somehow caused by human activities like fishing, which seems increasingly probable. Crown-of-thorns starfishes do not undergo such population explosions in the eastern Pacific, probably because of the limited development of coral reefs. But a very different scenario is likely if the starfish were introduced into the Caribbean, where the extent of reefs is much greater and overfishing is extreme.

Another species whose introduction would probably cause great ecological change is the unidentified pathogen that caused the epidemic mortality of the long-spined Caribbean sea urchin *Diadema antillarum* (see fig. 2–11). The eastern Pacific *Diadema mexicanum* is ex-

2–18. Pacific species whose introduction could dramatically change conditions in the Caribbean: (A) the crown-of-thorns starfish *Acanthaster planci*, seen here eating a large table *Acropora* on the Great Barrier Reef, is a voracious predator of reef corals throughout the Indo-Pacific. Photograph by John Ogden; (B) the venomous sea snake *Pelamis platurus*. Photograph by Carl Hansen.

tremely similar genetically and dies quickly in experiments when exposed to sick Caribbean relatives. Presumably the introduction of the pathogen to the eastern Pacific would cause similar mass mortality. Finally, introductions may be harmful for many reasons besides their ecological effects. For example, the venomous eastern Pacific sea snake *Pelamis platurus* (fig. 2–18B) causes only rare fatalities to unsuspecting fishermen who catch them accidentally by trawling. *Pelamis*'s danger could be blown out of all proportion, however, if it were accidentally introduced into the Caribbean, with unknown consequences for Caribbean tourism.

There is growing regional concern about the degradation of Caribbean coral reefs, sea grasses, and mangroves, and the economic consequences of such losses for subsistence and tourism. The world's first regional coral reef monitoring program, the Caribbean Cooperative Monitoring Program, was founded to establish baseline information on the status and trends of these communities throughout the Caribbean. Experimental marine reserves are being set up in Belize, Honduras, and Costa Rica to test the effects of stopping fishing on coral reefs and of protecting green turtles on the beaches where they lay their eggs. Preliminary results are encouraging, but there is increasing evidence that coral reefs and related ecosystems are closely linked throughout the region. For example, the floating larvae of corals and fishes are commonly transported hundreds of kilometers from their parents before they settle down, so the recruitment of a particular species in Honduras may depend on its reproduction in Colombia or Panama. Sustainable use will therefore require dramatic action at both local and regional levels.

CHAPTER THREE

Central American Landscapes

DAVID RAINS WALLACE

The seven nations that make up Central America have a combined area comparable to that of France or Texas. Yet the landscapes of Central America are extremely diverse; the area has long Pacific and Caribbean coastlines and many mountain ranges (figs. 3–1, 3–3, 3–4). Some 7 percent of the earth's species live here in less than 1/2 percent of the earth's land area. Geological and evolutionary complexity have given rise to this diversity. Landscapes of surpassing beauty are its most obvious effect.

The pioneer archaeologist John Lloyd Stephens wrote that "the highest pleasure" of his travels from Belize to Costa Rica in 1839–41 was "that derived from the extraordinary beauty of scenery, constantly changing." In spite of the great social and environmental transformations that have occurred since then, a traveler retracing his route could say the same thing today.

Central America is a miniature continent in the number and distinctiveness of its regions. On modern roads, it is possible to cross from the Caribbean to the Pacific in a few hours. Yet the traveler can witness many environments in that time: coral reefs, savannas, rain-forested foothills, ranges of cloud forest (watered by the mist from clouds), interior farming valleys, pine-forested volcanoes, semiarid Pacific lowlands. The overall impression is not of an attenuated land bridge between North and South America, but of a substantial landmass. Central America has three major lowland regions: the Petén, the Mosquitia, and the Nicaragua Depression; three major highland regions: the northern

Crystalline Highlands, the Pacific Volcanic Arc, and the Talamanca Massif; and a number of peripheral regions of fascinatingly varied origin and character.

Belize and the Petén

One of the most famous peripheral regions, and the northernmost, is the Belize barrier reef and the system of brackish and freshwater lagoons and wetlands inland from it (fig. 3–1). Lying on a succession of north-south trending limestone ridges, reefs and lagoons form a continuum: it is difficult to distinguish land from water when overflying them. The barrier reef is the longest continuous coral structure in the Western Hemisphere and the second longest in the world. Much of it is presently exposed as elongated cays, including Ambergris and Caulker cays (shown on figure 3–1), which are vegetated with mangroves, coconuts, or hardwood forest according to their topography and history.

Toward the reef's seaward side, tectonic activity has lowered the sea floor to depths of 300 meters and more. A few reef and cay complexes on this periphery, Lighthouse Reef, for example, have developed an atoll-like formation, with elongated rings of cays enclosing shallow lagoons standing in deeper waters. Belizean atolls, however, are not developed on extinct volcanoes like the classic Pacific atolls first analyzed by Darwin. Coral and fish populations around the atolls are spectacular, with a mingling of shallow- and deep-water forms.

Erosion by dissolution of the limestone substrate in this region has created a system of subterranean connections between sea and land. Freshwater springs flow into offshore reefs, and such reef fish as snappers appear in streams far inland, emerging from underwater caves.

Surface tidal flow connects the Caribbean with a chain of brackish lagoons directly inland. Surrounded by mangroves, the lagoons provide feeding and calving habitats for Belize's sizable manatee population, which also travels throughout the barrier reef. Inland from the tidal lagoons is a system of elongated, largely seasonal freshwater lagoons ringed with swamps of logwood and freshwater mangrove and connected by streams bordered with gallery forest. The lagoon complex has been called a Belizean Everglades and may in fact retain a greater abundance and diversity of waterbirds and other wildlife than its beleaguered Florida counterpart.

A savanna of scattered Caribbean pines (*P. caribacea*) and small palms and hardwoods occupies much of the pine ridges that separate the lagoons. Dry season fires probably maintain the savanna: in forest

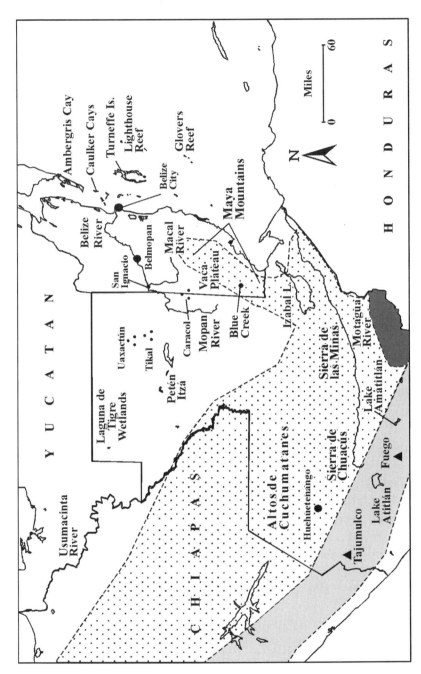

3–1. Map of Guatemala and Belize showing the location of the principal physiographic features as well as some modern and ancient cities mentioned in the text. The light gray area outlines the Volcanic Arc, the dotted area is the Crystalline Highlands, and the dark gray area delimits the High Volcanic Plateau.

reserves where fire is suppressed, it is succeeding to hardwood forest of oaks and various tropical species (fig. 3–2). Hardwood forest prevails along streams and in other favorable sites. In northern Belize's coastal plain, a relatively low forest similar to that of Yucatán grows.

Subsistence farming and fishing by such indigenous peoples as the Creole, Garífuna, and Maya remain important land uses in coastal Belize. Commercial farming is practiced mainly by Mennonite settlers concentrated in the San Ignacio and Blue Creek areas inland from the coastal plain (see fig. 3–1). Commercial fishing for conch, spiny lobster, and green turtle also occurs. Tourism has become a major land use factor, particularly on Ambergris and Caulker Cays in the north, although Belize has so far resisted the temptation of mass high-rise beach development, as practiced in Mexico.

West of the coastal plain, the land rises toward the escarpment that marks Belize's border with Guatemala. In the northern half of the country, where the coastal plain is broad, that rise is low and gradual. In the south, it is swift and dramatic because the granitic mass forming the Maya Mountains has risen through the limestone substrate to a height of 1120 meters (see fig. 3–1). Steeply eroded limestone cones rising from the coastal plain mark the range's foothills. A limestone tableland, the Vaca Plateau, is perched on the range's west side but drains to the north and east via the Macal, Mopán, and Belize rivers.

From north to south, a mature hardwood forest covers the escarpment. Only the scattered farmlands at San Ignacio and Blue Creek interrupt it. The forest approaches high rain forest conditions like those of northern Honduras in the south, where the dry season is shorter and the Maya Mountain peaks create lush watersheds. In the north it is drier and more seasonal but still impressive, with a great variety of tropical species.

Belize's western escarpment is the edge of the Petén, a rolling limestone plain that extends into Yucatán in the north and to the Guatemalan highlands in the west. The Petén is classic karst terrain, with few surface streams in its interior and extensive sinkhole and cave formations. The caves, particularly impressive in the Vaca Plateau, contain some of the longest subterranean chambers in the world. Sinkholes and related depressions called *bajadas* are the only dry season water sources in much of the region. Surface streams are mostly on the periphery, with drainage generally trending south or west toward the Usumacinta River system, except near the Belizean border. The Laguna del Tigre region in extreme northwest Guatemala is a vast complex of seasonal mudflats, sawgrass marshes, and palm or hardwood hammocks,

3-2. A view of the pine savanna in the Río Bravo Conservation Area of Belize and its interface with the tropical hardwood forest. Photograph by Kevin Schafer, 1996.

perhaps the largest single wetland remaining in Central America (see fig. 3–1).

Hardwood forest similar to Belize's still covers most of the northern Petén, notably within the Maya Biosphere Reserve. The area has not been logged as thoroughly as Belize and contains the largest commercial hardwoods such as mahogany (*Swietenia*) in the north of Honduras, although timber poaching via Belize is occurring. South of Lake Petén Itzá in the central Petén, shifting cultivation and cattle ranching have removed most primary forest within the past three decades. Regional population has increased threefold in that period by immigration from the highlands.

An important aspect of Petén and Belizean landscapes is the impact of Classic Maya civilization, which densely populated much of both regions for 2000 years until its mysterious collapse about 1100 years ago. The major city-states of Tikal, Uaxactún, and Caracol (see fig. 3–1) covered hundreds of square kilometers of limestone upland with workplaces and homes as well as with the giant ceremonial structures for which the region is famous. Smaller cities and other settlements existed on virtually every favorable site. These sites and the complicated system of reservoirs, storage caves, and raised bed, irrigated farmlands that supported them must have greatly altered the original topography and vegetation. Evidence of heavy lake siltation, soil erosion, and declining health in Classic Maya sites suggests that environmental deterioration may have played a part in the region's abandonment. Today, Maya sites are major tourist destinations; the Villaflores area of Guatemala and the San Ignacio area of Belize are important centers.

Belize and the Petén have a lowland subtropical climate, with rain possible throughout the year, but most precipitation occurs from May to October. Hurricanes are common during the late summer and fall season: Belize moved its capital inland to Belmopan after Hurricane Hattie devastated Belize City in 1961. The winter dry season is most pronounced toward the border with Yucatán and least toward the south. In the late 1980s and 1990s, increased rainfall in the central Petén raised the level of Lake Petén Itzá, inundating some tourism developments.

The Crystalline Highlands

West and south of the Petén, high mountains appear. These are Central America's geological core, an ancient complex of sedimentary, metamorphic, and igneous rocks extending from western Guatemala to northern Nicaragua (shown by the dotted pattern in figures 3–1, 3–3).

These rocks have been uplifted by more recent tectonic activity into a series of massive blocks and ridges dissected by river gorges and fault valleys. Tectonic activity continues strongly, most notably along the Motagua Fault, which cuts through southeastern Guatemala. In the earthquake of 1976 that killed more than 20,000 people around Guatemala City, lateral movement along the fault averaged 100 centimeters, with a maximum of 325 centimeters.

Large natural lakes have accumulated in the two largest fault depressions: Lake Izabal (see fig. 3–1) in Guatemala's Motagua Depression and Lake Yojoa (see fig. 3–3) in Honduras's Ulúa Depression.

The limestone Altos de Cuchumatanes Plateau (see fig. 3–1), just south of the border with the Mexican state of Chiapas, is the highest of the uplifted blocks, with a general elevation of more than 3000 meters. Discovery of *moraines* (sedimentary debris from glaciers) in shallow river valleys on the plateau in 1968 showed that it was glaciated during the Pleistocene. The plateau presently resembles the Mexican highlands, with a vegetation of scattered pines and montane scrub-grassland called *zacatal.* The steep slopes that drop into surrounding valleys support a pine-oak-laurel cloud forest, although much of this has been cleared. Some areas of serpentine bedrock occur along the fault lines at the base of the plateau.

Uplifted Mesozoic limestone also is a common bedrock in the Sierra de Chuacús and Sierra de las Minas ranges of central and southern Guatemala (see fig. 3–1) and in the adjacent cordilleras of Honduras. Granitic igneous and metamorphic rocks lie beneath the limestone and come to the surface to the south and east, where they give the name Crystalline Highlands to cordilleras that reach to the Caribbean Sea in Honduras and Nicaragua. Nombre de Dios (see fig. 3–3) forms a highly dramatic backdrop to the north coast of Honduras, rising to 2435 meters near La Ceiba. The cordillera's peaks extend underwater to the northeast, emerging as the Bay Islands. The Crystalline Highlands extend south through the central and southern Honduras cordilleras to northern Nicaragua's Cordillera Isabella (see fig. 3–3).

Because of their varied topography—ranging from more than 3000 meters to near sea level in such fault depressions as Guatemala's Motagua Valley and Honduras's Ulúa Valley—the Crystalline Highlands have a complex vegetation. Generally, eastern slopes exposed to the Caribbean support lush rain forest and cloud forest. Valleys just inland from lush coastal peaks can be semiarid because the prevailing winds have dropped their moisture during elevation and cooling over the coastal mountains—this is called the rain shadow effect. Those valleys

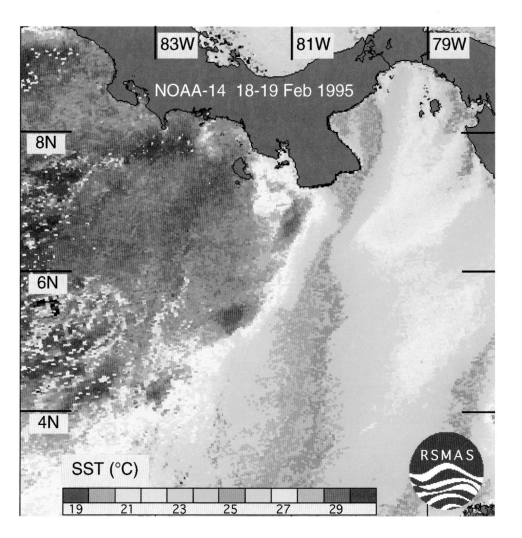

Plate 1. Composite satellite image of sea surface temperature showing the cold temperatures and intense upwelling in the Gulf of Panama compared with high temperatures and no upwelling in the Gulf of Chiriquí. Image produced by Guillermo Podestá, Joanie Splain, and Peter Glynn, University of Miami, Rosenstiel School.

Plate 2. Shallow coral reef communities dominated by branching species: (A) Caribbean reef crest community of elkhorn coral *Acropora palmata* in about 3 meters depth at Discovery Bay, Jamaica. Photograph by Nancy Knowlton; (B) Caribbean staghorn coral *Acropora cervicornis* community in about 15 meters depth at Discovery Bay. Photograph by Linton Land; (C) eastern Pacific *Pocillopora "damicornis"* community in the Gulf of Chiriquí. Photograph by Juan Maté; (D) same reef as (A) just after Hurricane Allen reduced the elkhorn community to rubble all along the north coast of Jamaica on 6 August 1980. Photograph by Linton Land.

Plate 3. Fishing in the productive waters of the Gulf of Panama: (A) single net haul of anchovies attracts numerous pelicans, whose reproduction in dense nesting colonies coincides with upwelling and peak anchovy production; (B) abundant anchovies; (C) collection of fish captured by a single trawl for shrimp; (D) the small basket of shrimp obtained from the same trawl. The remaining, mostly dead fish were thrown away. Photographs by Marcos Guerra.

Plate 4. Mangroves and sea grasses on the Caribbean coast of Panama: (A) turtle grass bed (foreground) and red mangrove in a back reef lagoon, Galeta, Panama. Photograph by Marcos Guerra; (B) red mangrove with prop roots. Photograph by Marcos Guerra; (C) crab, *Goniopsis cruentata,* feeding on mangrove seedling. Photograph by Norman Duke; (D) turtle grass bed with shaving brush calcareous alga *Penicillus* and white carbonate sediment derived from the skeletal fragments of algae and animals. Photograph by Marcos Guerra.

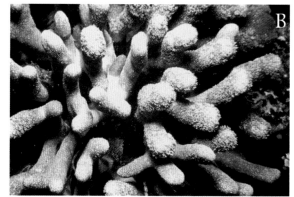

Plate 5. Coral polyps and various different colony forms produced by different budding patterns: (A) coral polyps showing tentacles with stinging cells (white dots) surrounding mouth. Photograph by Chrisse Harwanko; (B) *Porites,* with small, poorly defined polyps, forms distinctive branching colonies. Photograph by Ernesto Weil; (C) *Diploria,* a brain coral, has polyps that bud incompletely in one direction, creating meandering valleys at the surface of the massive dome-shaped colonies; (D) *Montastraea,* the star coral, also forms massive dome-shaped skeletons, but its polyps bud in all directions and separate completely. Photographs C and D by Carl C. Hansen.

Plate 6. Bleaching of the star coral *Montastraea* in western San Blas, north coast of Panama: (A) colony has lost virtually all its symbiotic algae so that the translucent living coral appears ghostly white from the color of its skeleton below; (B) partially bleached colony reflects the differences in type of algae that live in different parts of the host coral; (C) unbleached coral. Photographs by Ross Robertson.

Plate 7. In 1986, 30 million gallons of crude oil spilled near Bahía Las Minas, Caribbean coast of Panama. The photograph shows the oil washing ashore onto Galeta reef. This site had been studied for more than ten years before this spill and was monitored for more than six years afterward by the Smithsonian Tropical Research Institute. Photograph by Carl Hansen.

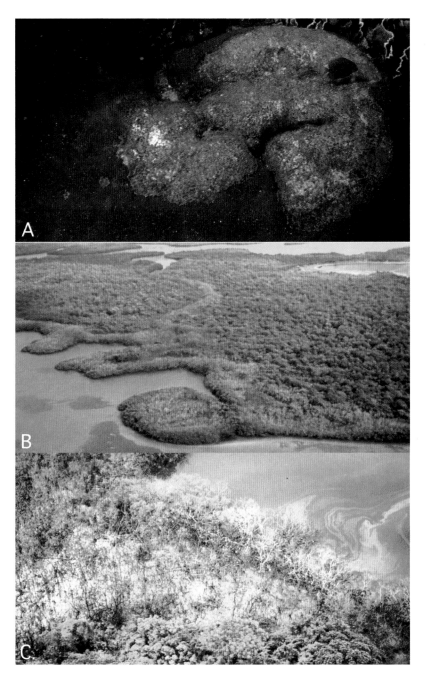

Plate 8. Short-term effects of the oil spill in Bahía Las Minas on the Caribbean coast of Panama: (A) colony of the coral *Siderastrea siderea* partially killed by the oil on its upper surface, which is now covered by algae and sediment; (B) partial view of the 27-kilometer-long zone of dead (gray) red mangrove trees 8 months after the oil spill; (C) oil washing out from dead red mangrove forest, as a result of heavy rains, 3 years after the oil spill. Oil was still being washed out in large quantities until another major spill in 1995 obscured the effects of the previous accident. Photographs by Carl Hansen.

Plate 9. (A) A jadeite mask from Chalchuapa, El Salvador; (B) "potbellied" statue from Santa Leticia, El Salvador. Photographs A and B by Payson Sheets; (C) fragment of an Olmec jadeite pendant from Salitrón Viejo, Honduras, latter half of the first milennium. Photograph by K. Hirth, of jade described in K. Hirth and S. Grant, "Ancient Currency: The Style and Use of Jade and Marble Carvings in Central Honduras," in *Pre-Columbian Jade,* ed. F. Lange (University of Utah Press, 1993), fig. 13.13; (D) jadeite pendant in Olmec or Izapan style from Tibas, Costa Rica, that may have been an heirloom, about A.D. 300. Photograph by Héctor Gamboa; (E) Macaracas polychrome plate from central Panama, about A.D. 700–1000. Photograph by Carl Hansen; (F) a human face ceramic mask from Miraflores, Bayano River, Panama, A.D 700–900. Photograph by Richard Cooke.

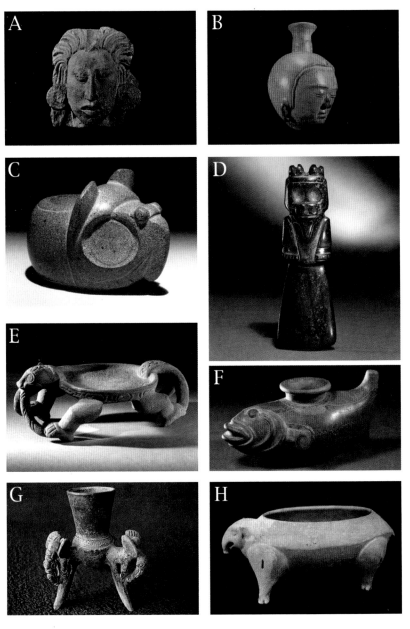

Plate 10. (A) Stone sculpture of the long-haired Maya maize god from Copán, Honduras. Photograph, American Museum of Natural History; (B) an Usulután human effigy vase from Chinandega, Nicaragua; (C) a stone mace head from Gran Nicoya, A.D. 500–1000. Photographs B and C from C. Baudez, *Ancient Civilization of Central America* (Barrie and Jenkins, London), pl. 7; (D) bird-axe jadeite ornament from Gran Nicoya, A.D. 500–1000; (E) jaguar-shaped stone stool from Caribbean watershed of central Costa Rica. Photographs D and E by Dirk Bakker; (F) a zoned bichrome fish effigy, 500 B.C.–A.D. 500. Photograph by Martin Matvig; (G) tripod vessel with bird legs from Caribbean watershed of central Costa Rica, A.D. 800–300. Photograph by Richard Cooke; (H) "Bisquit" terracotta tapir effigy from Gran Chiriquí, A.D. 1100–1520. Photograph by Carl Hansen.

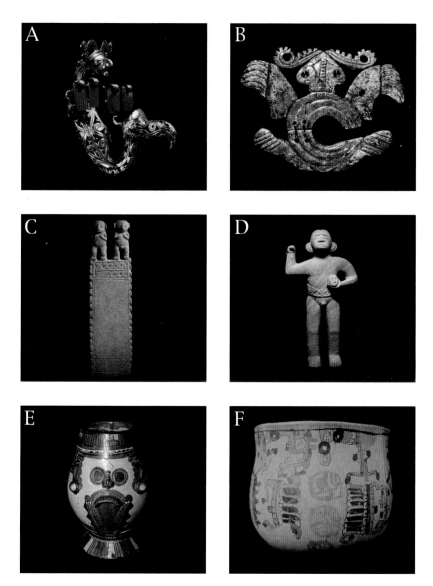

Plate 11. (A) Cast gold human-animal figure from Venado Beach, Panama, probably about A.D. 600–700; (B) *Spondylus* shell sea turtle nose-ring from Venado Beach, Panama, probably about A.D. 600–700. Photographs A and B by permission Dumbarton Oaks Museum, Washington, D.C.; (C) stone grave marker from central Costa Rica; (D) standing stone human figure holding an axe and human trophy head from central Costa Rica. Photographs C and D by Dirk Bakker; (E) polychrome vessel from Gran Nicoya with the face of Tlaloc, the Mexican rain-god, A.D. 1100–1350. Photograph by Martin Matvig; (F) Ulúa polychrome jar from the Sula valley, Honduras, A.D. 600–900. Photograph by George Hasemann.

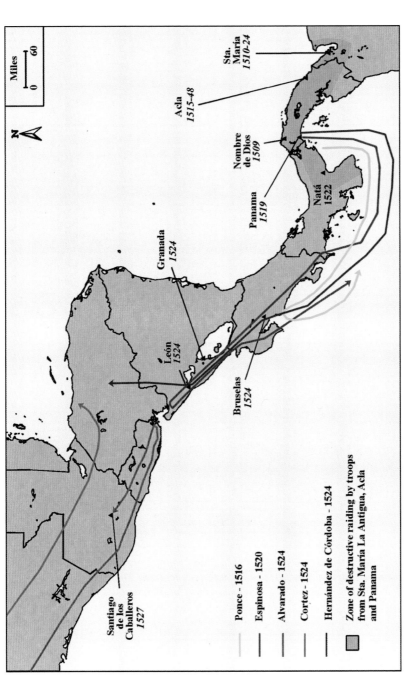

Plate 12. Map showing the routes of the conquest of Central America. The area in gray indicates the first region of Central America to be invaded by raiding parties from the early north coast settlements. Later, the extended expeditions set out from the city of Panama, and Alvarado and Cortez came down from the north. From Murdo J. MacLeod, *Spanish Central America, A Socioeconomic History*, 39, fig. 1.

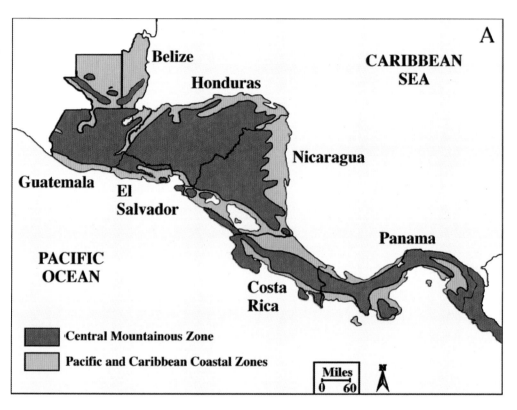

Plate 13. (A) Regional topographic divisions of Central America that have controlled land use and settlement patterns.

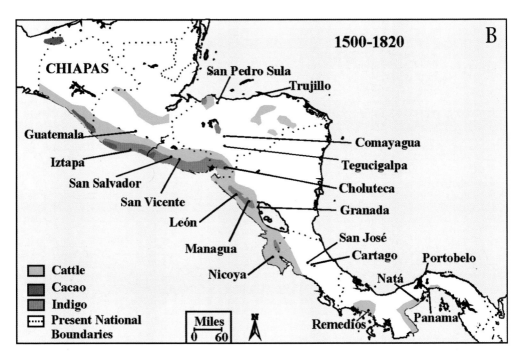

Plate 13. (B) Central America under Spain, 1500–1820, distribution of the main cities and economic activities.

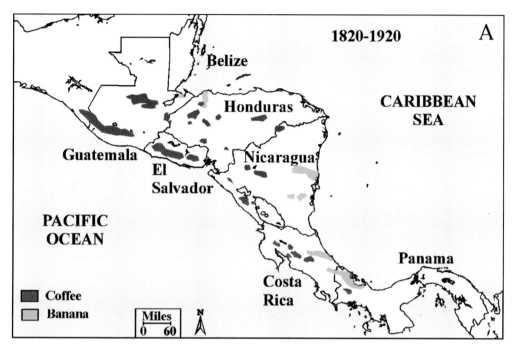

Plate 14. (A) Main areas of coffee and banana production in Central America, 1820–1920.

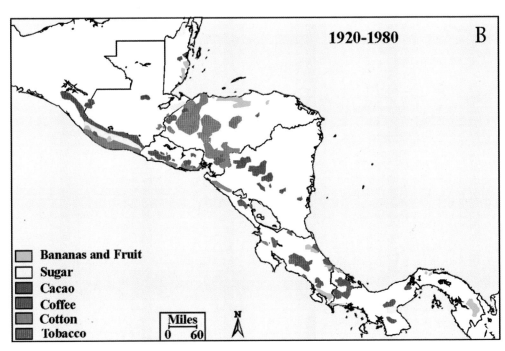

1920-1980

B

Bananas and Fruit
Sugar
Cacao
Coffee
Cotton
Tobacco

Miles
0   60

N

Plate 14. (B) Distribution of the principal export crops in Central America, 1920–80.

Plate 15. (A) Deforestation by slash-and-burn agriculture in the central highlands of Honduras.

Plate 15. (B) Removal of highland pine forest and subsequent overgrazing results in massive soil degradation and erosion. Photographs by Stanley Heckadon-Moreno.

Plate 16. (A) Major gullies erode soil to bedrock and wash millions of tons of soil from the central mountain ranges to the sea.

Plate 16. (B) Illiterate and impoverished campesino and his children stand in front of a Río Plátano Biosphere Reserve boundary sign that they cannot read. This symbolizes the dilemma of Central American conservation: how to protect and maintain the rich natural resources in the face of the desperate poverty and economic needs of the advancing colonists. Photographs by Stanley Heckadon-Moreno.

Plate 17. (A) Miskito women smoke fish along the Río Plátano in the Río Plátano Biosphere Reserve, Honduras.

Plate 17. (B) Garífuna women bake coconut bread in a village on the north coast of Honduras. Photographs by Peter Herlihy.

Plate 18. (A) Kuna woman sews a *mola* tapestry to be sold to tourists; (B) Emberá teenager is dressed in traditional manner with skin painted with black dye from jagua fruits (*Genipa americana*).

Plate 18. (C) Lahas Blancas, an Emberá village on the banks of the Chucunaque River in the autonomous Comarca Emberá/Wounaan, Darién, Panama. Photographs by Peter Herlihy.

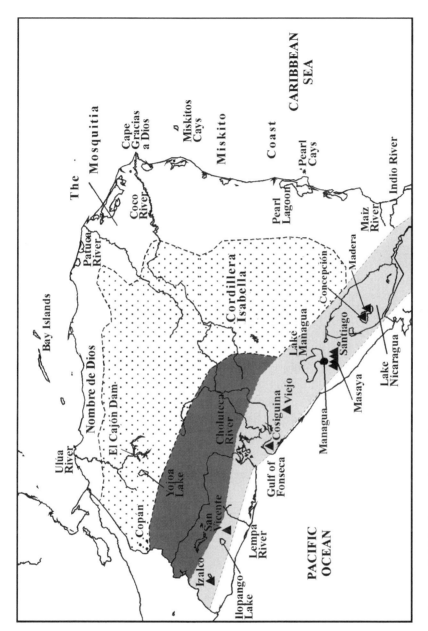

3-3. Map of El Salvador, Honduras, and Nicaragua showing the location of the principal physiographic features and some modern and ancient cities mentioned in the text (symbols and shading are as for figure 3–1).

now support a cactus-agave-acacia association on their floors, but this vegetation may be a degraded remnant of a more diverse, dry deciduous forest that existed before human settlement.

Above valley floors, interior mountains support successive altitudinal associations in which pines are often the predominant trees. Between 800 and 1600 meters, ocote pine (*P. oocarpa*) forms a sparse woodland with evergreen oaks and shrubs like *Miconia* on foothills and low plateaus. Pinabete pine (*P. maximinoi*) grows with sweet gum (*Liquidambar*) and other temperate deciduous genera between 1100 and 2000 meters. A band of evergreen cloud forest with an understory including tree ferns, begonias, and other seemingly incongruous "tropical" plants grows above the drier pinabete and sweet gum association. At the highest elevations, above about 2500 meters, a glacial relict association including fir, yew, and several other pines occurs.

Varied topography also conditions a diversity of climates, ranging from pluvial on Caribbean slopes, to warm temperate in interior valleys, to cool temperate on high ridges and plateaus. Night frost is common at higher elevations, particularly during the October to May dry season, but snowfall is rare even in the Guatemalan highlands. Although most precipitation occurs from May to October, rainfall is virtually a daily affair on Caribbean slopes, while mist and drizzle are permanent in the cloud forest belt. Interior valleys and foothills, on the other hand, may suffer months of drought in the dry season.

Because of rugged topography and relative soil infertility, the Crystalline Highlands have supported less human population than the volcanic highlands to the west. The Guatemalan and Honduran coasts of the Gulf of Honduras are exceptions to this, as are the Motagua and Ulúa Depressions. On the other hand, much of Nicaragua's and Honduras's cordilleras remains wilderness, and Guatemala's Sierra de las Minas is its greatest reserve of biotic diversity. Although occupied by numerous Maya groups for thousands of years, the Altos de Cuchumatanes and other plateaus are still rather sparsely settled compared to the volcanic region. In much of the Guatemalan highlands, subsistence farming and livestock raising by indigenous Maya groups remain major land uses.

A large tourist industry has grown up around living Maya culture, centered mainly in the volcanic highlands but with a center around Huehuetenango in the Crystalline Highlands as well (see fig. 3–1). Other tourism centers related to the Crystalline Highlands are the diving and snorkeling resorts of the Honduran north coast and Bay Islands. The archaeological Maya center at Copán (see fig. 3–3) is also important, and tourism in national parks in Honduras is growing.

Deforestation has been extensive in accessible areas throughout the region. In 1990, satellite photos of the Sierra de las Minas indicated that as much as half the range had been deforested in the previous decade. Watersheds around heavily populated areas such as the Comayagua Valley in Honduras have been reduced to moonscapes. Much of the exodus of small farmers that is spilling toward the Caribbean lowlands comes from deforested and denuded watersheds of the Crystalline Highlands.

The Crystalline Highlands have always been Central America's major mining region. Pre-Columbian jade may have been mined in Guatemala's serpentine deposits. Precious metal mining in the region was an impetus to Spanish colonization, and mining is still an important industry in the Honduran and Nicaraguan cordilleras. Environmental and social problems often accompany it: for example, the water pollution and labor disputes at the El Mochito lead mine on Lake Yojoa in the Ulúa Valley.

The Mosquitia Coast

East of the Crystalline Highlands, the Caribbean lowlands of eastern Honduras and northeastern Nicaragua, known as La Mosquitia in Honduras and as the Miskito Coast in Nicaragua, comprise the largest forested basin in the region, a kind of Central American Amazon (see fig. 3–3). A zone of subsidence since the early Cretaceous, the Mosquitia has accumulated about 4500 meters of sediments eroded from the Crystalline Highlands. The large rivers that wind through its low hills and valleys continue to deposit these sediments in swampy estuaries.

The Miskito Coast lacks a continuous barrier reef, but shallow waters extend far offshore, encompassing vast sea grass beds and scattered reefs. These provide perhaps the major habitat in Central America for economically valuable marine wildlife, especially green turtles, spiny lobsters, and conchs. Recent observations suggest that the manatee population may rival that of Belize in significance. The largest reef complexes are the Miskito and Pearl Cays (see fig. 3–3).

A number of large tidal lagoons, separated by sandbars from the sea, line the shore in Honduras and Nicaragua. Mangrove swamp prevails around the lagoons and along the meandering channels that connect them. Freshwater swamps and marshes extend great distances inland from the lagoons, and there are some freshwater lagoons.

A Caribbean pine savanna similar to that of Belize occupies vast areas of uplands, with typical tropical savanna tree species such as nancite (*Byrsonima crassifolia*) and calabash (*Crescentia*) growing among the pines. It is unclear why savanna is so prevalent in this thinly populated region, which has a climate capable of supporting rain forest. Fire is undoubtedly a factor; infertility of heavily leached, residual iron-oxide-rich (lateritic) soils probably is another. Hardwood forest occurs along streams and in other alluvial sites in the savanna, and rain forest prevails farther inland, where the Crystalline Highland foothills rise.

Recent archaeological excavations have found evidence of significant pre-Columbian urbanization in the Río Plátano drainage of Honduras. These so-called white cities are thought to date from the four centuries before the conquest and may have been related to commercial production of cacao for the Mesoamerican empires. The Mosquitia was never settled in the wholesale manner of the Petén, however. Today, the Miskito and Garífuna peoples populate the coast relatively densely in scattered areas, while the Pech and Sumu have more sparse populations inland. Mestizo immigration from the west has been extensive. Subsistence farming on floodplains, fishing, and hunting remain significant, interrupted sporadically by various activities that exploit local resources with outside financing. The latest of these are a spiny lobster fishery and an attempt to log savanna pines for pulpwood.

The Mosquitia has been characterized as having two seasons, the rainy season and the monsoon. Flooding renders a large part of the region inaccessible for much of the year. A November-April dry season exists, but major storms may arrive at any time — northeasters in winter, hurricanes in fall. Full-scale hurricanes do not usually reach south of Nicaragua, but storms often cause major flooding along the Caribbean coast of Costa Rica and Panama.

The Nicaragua Depression and Pacific Lowlands

In southeast Nicaragua, the Miskito Coast merges with Central America's other great lowland, the Nicaragua Depression (figs. 3–3, 3–4), which cuts through the land like a 50-kilometer-wide trough from the San Juan River on the Caribbean coast to the Gulf of Fonseca on the Pacific, a distance of about 500 kilometers. It is nowhere more than 50 meters above sea level, and half of it is covered with freshwater lakes, the most important of which are Managua and Nicaragua. Lake Nicaragua is the largest body of freshwater between the Great Lakes and Lake Titicaca. The depression clearly marks the border between north-

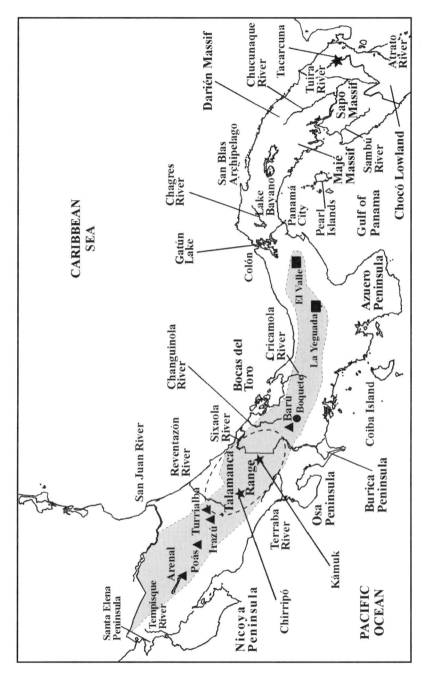

3–4. Map of Costa Rica and Panama showing the location of the principal physiographic features (shading as for figure 3–1).

ern Central America's ancient, continental rocks and the more recently emerged oceanic rocks of southern Central America.

The depression has been a zone of subsidence and sedimentation for most of the Tertiary. Sediments are volcanic as well as alluvial, and there is evidence of recent volcanic activity even on the Caribbean coast. The lakes are underlain by terrestrial sediments and probably formed fairly recently as a result of faulting activity. They are relatively shallow, averaging about 8 meters in depth.

The presence of such marine fauna as sharks (*Carcharhinus leucas*) and sawfish (*Pristis perotteti*) in the lakes, once thought to indicate a former seaway, has been shown to result from migration up the San Juan River from the Caribbean. Lake Nicaragua is still an important fishery for *guapotes* (*Cichlasoma*), gars (*Atractosteus tropicus*), and other locally consumed food fishes. Smaller lakes, including Lake Managua, are polluted with pesticides and other chemicals.

Perhaps because of volcanic substrates, the vegetation in the Nicaragua Depression is somewhat richer than in the Mosquitia. The watersheds of the Indio and Maíz rivers (see fig. 3–3) north of the San Juan River support a rain forest of Amazonian proportions, and the Llanura de Tortuguero south of the San Juan is comparable (fig. 3–5). There are also large areas of *Raphia* palm swamp and mangrove. Early accounts such as Thomas Belt's "The Naturalist in Nicaragua" indicate that high rain forest extended inland as far as Lake Nicaragua. Today, most primary forest inland from El Castillo at the westernmost of the San Juan's four rapids has been replaced by farms and ranches. West of Lake Nicaragua, the dry deciduous forest typical of northern Central America's Pacific coast replaces evergreen forest. In the depression, this largely had been replaced with savanna as early as circa 1800, the time Belt was writing. Today, only fragments remain on the volcanic escarpments bordering the depression, on volcanoes, and on islands in the lakes and in the Gulf of Fonseca. More significant, dry forest remnants persist in Costa Rica's Guanacaste Province (fig. 3–6), south of the depression, where lowlands and volcanic plateaus surrounding the Tempisque River basin have remained less heavily populated.

Most dry deciduous forest trees lose their leaves during the largely rainless October to May of the Pacific lowlands. This gives the landscape a gray, bare aspect, incongruously like temperate forest in winter, although the dry heat belies this. Many tree species, including mayflowers (*Tabebuia*) and *pochotes* (*Bombacopsis quinatum*), also belie the wintry aspect by producing flowers and fruit in the leaf dormant period. The deciduous forest occurs from sea level to about 1000 meters,

3–5. A view of the Caribbean coastal rain forest in Tortuguero National Park, Costa Rica. Photograph by Charles Luthin, Lighthawk.

3–6. A panorama of the freshwater marshes with bordering seasonal deciduous forest, typical of the Pacific coastal region, in Palo Verde National Park, Costa Rica. Photograph by Charles Luthin, Lighthawk.

where increased precipitation replaces it with various evergreen forest types. In the rainy season, the deciduous forest is indistinguishable from evergreen forest. Many plants and animals occur in both. Because of their long dry season, Pacific lowlands are generally perceived as the hottest part of Central America. Moisture-laden offshore trade winds cool the Caribbean through the dry months, but when the same winds reach the Pacific, they are dry and furnacelike and prevent ocean winds from moving east to cool the land. Yet seasonal dryness combined with adequate annual precipitation has made the Pacific lowlands more suitable for human occupation than the Caribbean.

Pre-Columbian agricultural societies in the western Nicaragua Depression and adjacent lowlands were so affluent that conquistadors called the region Mahomet's Paradise. Intensive farming remains the principal land use. Maize and fruit were the original crops, but cotton and sugar cane monocultures have prevailed in recent years, although because of environmental and economic problems attempts are being made to return to more diversified crops. Pesticide use has been profuse, with consequent declines in wildlife and many human deaths.

Mangrove swamps are the main remaining wildlands in the western depression and in much of the Pacific lowlands. The estuary of the Real River on the Gulf of Fonseca in northwest Nicaragua is one of the largest of these, and others occur around the gulf, helping to support a shrimp fishery. Overcutting for charcoal and pesticide pollution threaten the ecology of the mangrove.

The Volcanic Highlands

Although the Nicaragua Depression clearly demarcates northern from southern Central America, it shares one distinctive and awe-inspiring feature with both—volcanoes. The chain of active and recently active volcanoes that extends from Tajumulco in western Guatemala to Barú in western Panama is the spine of Central America's most important region (see light gray areas in figs. 3–1, 3–3, 3–4). There are more than 350 volcanoes, and at least 20 are considered active. Deep ash deposits lie everywhere in the region and extend far out into the Pacific as well. Their effects on Central America's life, particularly its human life, have been both a curse and a blessing.

On the curse side, volcanic eruptions and explosions and related earthquakes have made life precarious. An explosion of Ilopango volcano (see fig. 3–3) in El Salvador about 1800 years ago drove the area's entire population away for centuries. In 1835, the explosion of Cosiguina (see fig. 3–3) in northwest Nicaragua was the greatest in recorded history in Central America and reduced the mountain from 2000 to 807 meters. Eruptions and earthquakes have destroyed the region's main cities so repeatedly that little colonial architecture remains. Much of Managua is still in ruins from the earthquake of 1972.

On the blessing side, volcanic ash and lava make a fertile, nutrient-rich soil, and, particularly since the inception of coffee growing in the mid-nineteenth century, the volcanic region has been the most prosperous and populous in Central America, although population pressure is now threatening that prosperity. Settlement historically has been concentrated in the valleys and plateaus below the volcanoes but now has spread upward as pressure has grown, and frontiers between cultivated land and remaining forest are visible on most peaks.

In spite of having related origins, Central America's Quaternary volcanoes differ from north to south. Guatemala's are the highest, having grown on the plateau about 70 kilometers inland from the present Pacific shore. Tajumulco (see fig. 3–1) near the Chiapas border is Central America's highest point at 4220 meters. A number of others are higher than 4000 meters. The volcanoes that existed where the calderas and lakes of Atitlán and Amatitlán (see fig. 3–1) now stand may have been even larger. Lake Atitlán is the largest caldera lake in Central America. Resulting from subterranean withdrawal of magma, the lake is very deep, draining to the Pacific by underground fault fissures. Drainage changes after the earthquake of 1976 reduced its level significantly.

Guatemala's volcanoes are also the most active. Fuego (see fig. 3–1), in central Guatemala near Antigua, has erupted at least 50 times since 1524. An eruption in 1971 deposited an ash layer a centimeter thick 160 kilometers from the crater and filled ravines below the crater 20 meters deep with glowing avalanches. Except for Arenal in Costa Rica, central Guatemala's volcanoes are the only ones to produce the spectacular fiery displays called *nuées ardentes,* or glowing clouds, described in more detail in chapter 2.

El Salvador's volcanoes are closer to the coast and lower than Guatemala's, most summits being around 2000 meters. They are almost as active, producing explosive eruptions and large lava flows. Izalco (see fig. 3–3), northwest of San Salvador, is famous for growing from a smoking fissure in a cornfield in 1770 to the present 1910-meter cinder cone. It erupted almost continuously until 1966. Two caldera lakes, Coatepeque to the west of San Salvador and Ilopango to the east, exist today.

Growing from the depression lowland, Nicaragua's volcanoes are the lowest in Central America. Viejo (see fig. 3–3) is the highest at a relatively low 1780 meters, but it is enormously impressive as it looms inexorably above its flat surroundings, often crowned with smoke and clouds; it seems snow-capped, although even Tajumulco is too low for snow. The twin volcanoes of Madera and Concepción on the Island of Ometepe in Lake Nicaragua (see fig. 3–3) are even more imposing, standing above the wind-driven waters. Nicaragua's volcanoes also produce explosive eruptions and heavy lava flows. The evidence of this is easily visible at Masaya volcano just south of Managua (see fig. 3–3). Lava flows from eruptions in 1670 and 1772 surround the five craters at the top of the 550-meter complex. Boulder-sized "bombs" ejected during eruptions lie about. An open well with molten lava at its base is located in the most active crater, Santiago.

Except for Arenal, Costa Rica's volcanoes have not produced lava flows in historical times, instead erupting steam and ash from deep craters. Rather than standing singly, they are clumped into massive cordilleras. The Guanacaste cordillera in northwest Costa Rica is comparable to Salvadorian volcanoes in height, with a few summits more than 2000 meters. Rincón de la Vieja in the center of the cordillera has had three prolonged eruptions since 1940 and has extensive hot springs and other geothermal features in the mixture of dry and evergreen forest at its foot.

Southeast of the cordillera, 1633-meter Arenal volcano (see fig. 3–4) is an anomaly, an isolated cone that was thought extinct until it erupted in 1968 with violent earthquakes and explosions that killed 78 people. Activity has continued into the present, with glowing clouds, lava discharges, and subterranean rumblings that can be heard many kilometers away.

The part of the Central Cordillera that rises north of Costa Rica's heavily populated Valle Central is the tallest volcanic structure south of Guatemala, crowned by 3432-meter Irazú and 3328-meter Turrialba (see fig. 3–4). The cordillera slopes gently to the roughly 1000-meter high Valle Central but drops sharply to the Caribbean lowland on its north side. Craters are relatively inconspicuous among the broad ridgelines and saddles but quite active nonetheless. Poás's 1.5-kilometer-wide crater produced an ash cloud 8000 meters high in 1910, and Irazú erupted large quantities of ash during 1962–66.

The continuous volcanic chain is interrupted south of Costa Rica's Central Cordillera by the nonvolcanic Talamanca massif (see fig. 3–4). Where the Talamancas end in western Panama, however, volcanoes reappear dramatically with 3475-meter Barú in Chiriquí Province.

Smaller volcanic peaks are scattered along western Panama's Central Cordillera as far as the Panama Canal basin. Panama's volcanoes have not been active in the past few centuries.

The chain of Quaternary volcanoes with its ash-enriched valleys and Pacific lowlands is not the only volcanic region in Central America. Inland from the active zone in southern Guatemala, western Honduras and Nicaragua, and northern El Salvador is a large, heavily dissected plateau formed from the remains of long-extinct Miocene and Pliocene volcanoes (see dark gray areas in figs. 3–1, 3–3). A relatively low plain when the volcanoes were active, the region has been uplifted during the Pleistocene into a rugged terrain of river gorges and gullied valleys.

Like those of the Crystalline Highlands, the vegetation and climate of the High Volcanic Plateau vary widely according to altitude and exposure. Altitudinal vegetation patterns on volcanoes north of the Nicaragua Depression resemble those of the interior Crystalline Highlands, ranging down from a shrub and grassland *zacatal* on Tajumulco above 3500 meters to a fir-yew-pine association above 2500 meters, to successive bands of cloud forest, pinabete–sweet gum forest, ocote pine–oak forest, and dry deciduous forest or acacia-cactus scrub. On the hills and valleys above the Pacific slope in Guatemala and El Salvador, volcanic soils and ample precipitation between 500 and 1500 meters supported a rich, diverse broadleaf evergreen forest with many endemic species. Almost all of this has been converted to towns or farmland, however, leaving only relics like the little-studied El Impossible forest in northwest El Salvador. Other low and mid elevation forest types have been largely replaced with sugar cane and cotton in the valleys and coffee in the hills, although El Salvador retains significant areas of traditional coffee plantations, where some forest has been retained to shade the coffee bushes.

In addition to intensive farming, cattle ranching and subsistence farming are important in the volcanic region. Guatemala has a major tourist industry centered in Antigua and Lake Atitlán. El Salvador's and Nicaragua's civil wars discouraged tourism until recently, but Costa Rica's tourism industry, centered on its system of national parks and other nature reserves, is the fastest-growing in Central America and recently surpassed coffee and bananas as a source of foreign revenue. Tourism is dispersed almost throughout Costa Rica, but the volcanoes of the Valle Central and Guanacaste are important centers. Panama too has an important tourism center in the Boquete area of Chiriquí Province (see fig. 3–4), under Barú volcano.

The Nicaraguan volcanoes mark the limit of much that is characteristic of northern Central America. From the plain below Viejo volcano, one sees dark, shaggy trees growing in the cone's ravines. These are ocote pines, the southernmost naturally occurring members of their species. The Caribbean pines of Mosquitia savannas also have their natural limit at the Nicaragua Depression. Many other northern tree species stop at the depression—fir, yew, *liquidambar,* hornbeam. Others—oaks, magnolias, alders—continue well into South America.

The absence of northern conifers makes altitudinal plant zonation less conspicuous (for the nonbotanist, at least) in the Costa Rican cordilleras than in Guatemala or Honduras. Moving east from the dry deciduous forested Guanacaste lowlands, one passes through a foothill zone in which a live oak (*Quercus oleoides*), closely related to *Q. virginiana* of the southeastern United States, is predominant, then into a highly diverse but little-studied tropical evergreen forest, followed by a diverse cloud forest with many tropical species and a less diverse cloud forest where oaks prevail. Forest occurs up to the summits of the Guanacaste Cordillera, where volcanic activity has not prevented its growth. Higher elevations of Irazú and Turrialba in the Central Cordillera have limited areas of heath and grassland related to larger areas of the same type on the Talamanca summits to the south. On their eastward slopes, Costa Rica's cordilleras receive lavish Caribbean precipitation and, because most of the Caribbean lowland forest has been cleared, support the richest rain forest remaining in the country. Vegetational patterns on the volcanic cordilleras of western Panama resemble those of Costa Rica.

Climate in the volcanic chain varies enormously. Temperatures can be surprisingly cold even at midelevations during the windy dry season. Frost is prevalent at higher elevations. Low elevation temperatures on the Pacific side can be infernal in the dry season.

Talamanca and the Pacific Peninsulas

Central America's third major highland region, the Talamanca Massif, is its smallest but in some ways its most dramatic (see fig. 3–4). The range rises like a wall from the Valle Central in the north, reaching a height of 3819 meters at Cerro Chirripó (*cerro* is a rounded hill or peak). It extends southeast about 200 kilometers to Cerro Punta, just north of Barú volcano. The Caribbean slope is gentler but dissected by rugged river gorges. Talamanca's geological origins and structure are elaborated in chapter 1.

Like the Altos de Cuchumatanes, the Talamancas bear witness to

Pleistocene glaciation, but the evidence is much more striking in the steep, granitic Talamancas. Cerro Chirripó and surrounding peaks are classic glacial mountain landscapes, with cirques (high, circular, ice-formed basins) and tarns (small lakes within a cirque) above U-shaped valleys and large moraines. The shrub and grassland vegetation of glaciated areas is called *páramo* after the similar vegetation of the Andes. The vegetation does have affinities with Andean páramo, although the Talamanca peaks may not be high enough to exclude forest under the present climate. Wildfires have been frequent in historical times: it is possible that fires have perpetuated a páramo vegetation lingering since the Pleistocene.

Costa Rican and Panamanian cordilleras and the Talamancas remain the largest island of undisturbed highland watersheds and forests in southern Central America. Most of Costa Rica's and western Panama's major rivers have their sources there.

The Talamancas are also an island of indigenous land uses. Except for banana, coffee and, pineapple plantations and vegetable farming in peripheral valleys, traditional subsistence hunting, gathering, and gardening still predominate among the Bribri, Cabécar, Teribe, and Guaymí peoples. Indigenous groups in the interior Talamancas still resist assimilation.

Central America's Pacific coast, south of the Nicaragua Depression (see fig. 3–4), is different from its north coast (see fig. 3–3). The north is almost a straight line of beaches, interrupted only by mangrove estuaries, from the Chiapas border to the Santa Elena Peninsula. The south is a complex of gulfs, bays, and peninsulas. Such peninsulas as Santa Elena, Nicoya, Osa, Burica, and Azuero differ markedly from the rest of Central America in their geology, topography, and vegetation.

The Santa Elena Peninsula (see fig. 3–4) is largely an elongate, heavily eroded mass of peridotite. Its predominant vegetation is a savanna of native bunch grasses, small trees like nancite (*Byrsonima crassifolia*) and rough-leaf tree (*Curatella americana*), and herbs and shrubs, some of which are endemic. This pattern of vegetation is common on peridotite-serpentine areas worldwide.

The Nicoya and Osa Peninsulas are composed of sedimentary and volcanic rocks of Cretaceous origin. Limestones have been eroded into caves in the Barra Honda area of the Nicoya Peninsula. The peninsulas enclose the Gulf of Nicoya and Golfo Dulce on their southeast sides.

The Nicoya Peninsula (see fig. 3–4) is steep and eroded at its margins; its interior is a rolling tableland 200–1000 meters high. The peninsula marks a transition between the dry deciduous forest of Guanacaste and the largely evergreen coastal forest of southwest Costa Rica. Remains of primary forest at Cabo Blanco at the tip show a blending of dry and evergreen forest characteristics. Most forest on the peninsula has been removed for crops and cattle ranching since the 1950s. The Tempisque basin to the east is one of Costa Rica's main farming areas but retains extensive freshwater wetlands (see fig. 3–6) along the lower Tempisque River.

The Osa Peninsula (see fig. 3–4) has a wetter climate than the Nicoya because it is shielded by the Talamancas from the winter trade winds so that precipitation arrives from the Pacific. It thus supported (and still supports in protected areas) the richest tropical lowland forest on Central America's west coast. Coastal uplands display a rain forest of Amazonian height and diversity: 500 tree species have been counted inside the borders of Corcovado National Park. Cloud forest and montane forest occur on interior highlands. A large lagoon and wetland complex is located in the west-center of the peninsula. The geologically similar Burica Peninsula (see fig. 3–4) across the Golfo Dulce in Panama retains much less forest than the Osa.

The Azuero Peninsula in Panama is the largest and geologically most complex of the Pacific peninsulas. Climate is drier than in the Osa, and dry deciduous forest occurs along with evergreen forest types. The peninsula's lowlands have been heavily populated since pre-Columbian times, and much land is deforested, with some areas reduced to semidesert conditions of sparse savanna-grassland and scrub. Archaeological and historical evidence suggests that deforestation was common in the preconquest period. Evergreen forest remains in the Cerro Hoya highlands in the southwestern part of the peninsula.

The Pacific slope of western Panama generally has seen prolonged deforestation, while the Caribbean slope of the Cordillera Central and the Bocas del Toro archipelago have retained large areas of rain forest. Most farm emigration from the deforested Pacific has been eastward to Darién because access to this area (on the Inter-American Highway) is easier than it is to the north. Bocas del Toro's islands have special biological interest because endemic populations apparently have evolved in the 6000 to 8000 years since rising post-Pleistocene sea level isolated them. The many endemic color phases of the poison dart frog *Dendrobates pumilio,* which is dark blue and red in its wide mainland range, are the best known examples of such evolution.

The Canal Zone and Darién

Panama's Cordillera Central terminates at the canal basin, a roughly 30-kilometer-wide lowland with a maximum altitude of 200 meters. The canal cuts across the land (which trends southwest to northeast at this point) in a northwest to southeast direction. This location tends to play havoc with North Americans' notions of compass directions because the Caribbean port of Colón is on the *west* coast and the Pacific port of Panama City is on the *east* coast (see fig. 3–4).

The Canal Depression marks the eastern limit of the strenuous tectonic activity that has produced the active Central American volcanic chain. The substrate's complexity and instability resulted in frequent, devastating landslides during construction of the canal. The canal has greatly influenced the depression's present landscape. Originally, the main drainage was to the Caribbean via the Chagres River, but the Chagres has been impounded at Lake Alajuela on its upper drainage and at Gatún on its lower and now contributes its flow to the canal's lock and dam system. Deforestation of the river's watershed, already under way before canal construction, threatens the water supply, and this threat has been the main impetus to conservation measures undertaken since the Canal Treaty of 1977.

The original forest in the depression was pluvial rain forest with swamps on the Caribbean side and more seasonal forest on the Pacific. Little primary forest remains within sight of the canal, but secondary forest has begun to reach considerable size, giving the landscape a sylvan aspect very different from the bare dirt expanses visible during canal construction almost a century ago.

At the canal's Pacific terminus, Panama City looks out on the Bay (and beyond that the Gulf) of Panama, one of Central America's most impressive seascapes. Bordered by the Azuero Peninsula on the west and by southern Darién on the east, the Gulf of Panama amounts to a small, rather shallow sea, with exaggerated tidal patterns. Up to 8 meters may separate the high and low tidemarks, and vast areas of mudflat and rock pool are exposed at low tide. The gulf has been one of Central America's major fisheries, but overexploitation has greatly reduced its productivity. Trawlers leaving behind long clouds of silt are a common sight from overflying planes.

Deforestation attendant on river impoundment is the contemporary reality of the Chepo Basin in western Darién east of the canal basin. The enormous Bayano Reservoir now takes up much of the basin, which extends between the San Blas Cordillera in the north and the

Majé Massif in the south (see fig. 3–4). A very low divide separates the Chepo Basin from the Chucunaque and Tuira basins of eastern Darién, which also have undergone extensive deforestation since 1970. All three basins drain into the Gulf of Panama. The San Blas and Darién cordilleras that border the basins on the Caribbean side and the Majé and Sapo cordilleras on the Bay of Panama side are small fault block mountains (see fig. 3–4). Their highest point is 1875-meter Cerro Tacarcuna near the Colombian border, but much of their extent is below 1000 meters. Only short streams drain into the Caribbean from the north side of the San Blas and Darién cordilleras.

Darién's remaining primary forest, located largely on the cordilleras, is the most "South American" in Central America, as might be expected. Huge emergent trees such as *cuipo* (*Cavanillesia platanifolia*), *barrigón* (*Pseudobombax septenatum*), and silkworm *(Ceiba pentandia)* rise above the subcanopy of *almendro* (*Dipteryx panamensis*), *rosa del monte* (*Browne macrophylla*), and hundreds of other species. Even the cloud forest of Cerro Tacarcuna and other peaks is dominated by a South American oak species (*Quercus humboldtii*), which evidently diverged from a Central American ancestor in Colombia, then spread north. As elsewhere, Darién forest grades from a pluvial form on the Caribbean to more seasonal forms in the interior and Pacific side. The April to January rainy season is intense even on the Pacific side, however, and widespread flooding makes overland travel impractical at that time.

Indigenous land uses persist mainly on the peripheries of Darién. The Kuna use the coral islets of the San Blas Archipelago for fishing and coconut plantations and grow crops by shifting cultivation on the mainland. The Emberá and Wounaan of eastern Darién are subsistence farmers and occasional hunter-gatherers. The usual Latin American colonization pattern of logging and shifting cultivation followed by cattle ranching has prevailed in the large basins.

Borders and Perspectives

Panama's eastern border follows the crest of the escarpment dividing the Tuira and Balsas basins from Colombia's huge, swampy Chocó Basin (see fig. 3–4). Drained by the Atrato River, the Chocó is geologically part of Central America in that it was the last piece of land to emerge during the isthmus's closing some 3 million years ago. Vegetation and climate also are quite similar to eastern Darién's, although the Chocó generally is regarded as one of South America's characteristic ecological regions.

Central America's northern limit is also somewhat vague. Chiapas and the Yucatán Peninsula resemble Guatemala and Belize in their geology and vegetation. Mexico and Colombia, however, are respectively such overwhelmingly North American and South American countries that it is simpler to identify Central America by its political boundaries than by rigid geological or biotic limits. Chiapas, Yucatán, and the Chocó are similar enough to Guatemala, Belize, and Darién, respectively, to describe them as parts of the same regions, with Chiapas included in the volcanic and Crystalline Highlands, and Yucatán as a continuation of the Petén's and Belize's limestone topography.

Describing Central America's many regions gives some idea of its diversity, but there are larger categories as well that provide unifying perspectives. Guatemala's Pacific side, for example, is in many ways more like Panama's Pacific side than it is like Guatemala's Caribbean side. Pacific shores may be straight and sandy in the north and crooked and rocky in the south, but they all have strong tides, offshore currents, and upwellings, in addition to waters unfavorable to coral reef formation. Caribbean shores may be sandy or rocky, but generally they have weaker tides and clear, shallow waters favorable to coral reefs. Caribbean shores as a rule get more than 3000 millimeters of rainfall a year, whereas Pacific shores get less than 2000. With a few exceptions, Pacific shores are drier, hotter, and more populous than Caribbean shores.

Altitude is another unifying category: Central America's highlands have similar qualities from Tajumulco to Barú. The climate above 1000 meters is one of the healthiest and most pleasant on earth, a permanent springtime of fresh nights and balmy days, with most rain falling in intense but short afternoon and evening cloudbursts. Highland vegetation differs sharply in the north and south, but in both there is an attractive mixture of tropical luxuriance and temperate crispness, with an interesting but not overwhelming diversity of species. Even in the relatively small extent of land above 2000 meters, daytime temperatures are comfortably warm for most of the year, though nights tend to be chilly.

Central America's lowlands have equally unifying qualities. The beauty and fascination of lowland forest are legendary, but other aspects, equally legendary, are less reassuring. Whether in Darién, Guanacaste, or the Petén, one encounters fierce heat, violent storms, and an almost incomprehensibly rich biodiversity in which every passing plague, from cholera to killer bees, seems to find a niche. Although diseases and parasites are less virulent than in the Old World tropics, such relatively harmless organisms as chiggers and botflies can make life miserable. Almost continuous rainstorms may persist for weeks or even months. Civilization has always rested uneasy in the lowlands, from the Classic Maya to the present cosmopolitan culture, and will continue to do so.

Of course, the best-known categories by which Central America can be both unified and divided are north and south—the north and south not only of Central, but of North and South America. "Continental" Central America from Chiapas to northern Nicaragua is a unit of sorts, as is "oceanic" Central America from southern Nicaragua to eastern Panama. Neither of these units escapes the influence of the continents to north and south. The North American influence is most conspicuous north of the Nicaragua Depression, where much of the mountain landscape might be in California, with its ravens, Steller's jays, and pine woods, but it continues south to the oak woods of Darién. The South American influence is most conspicuous south of the Nicaragua Depression, where much of the lowland landscape might be in Amazonia, with its great currasows, toucans, and rain forest, but it continues north to the Petén. The interpenetration of those influences throughout Central America is a reflection of the region's being the only functioning land bridge between two continents on the planet today.

CHAPTER FOUR

The Great American Faunal Interchange

S. DAVID WEBB

The formation of the Isthmus of Panama created a land bridge between North and South America that allowed mass migration of animals and plants through the tropics and into temperate latitudes both from north to south and from south to north. Because the two American continents had been widely separated for many tens of millions of years, their connection had a revolutionary impact. This extraordinary biological episode is known as the Great American Biotic Interchange (hereafter referred to as the interchange), and the drama began in Central America less than 3 million years ago.

The Ice Age, which began after the formation of the land bridge and was partly caused by it, has followed a series of regular cycles of cooling (glacials) and warming (interglacials). The glacial rhythms of the ice age, beginning about 2.4 million years ago and continuing today, have profoundly affected the course of evolution and extinction among the animals that crossed—and clashed—in the interchange. This chapter considers the role of Central America in radically changing the history of land animals in North and South America after the emplacement of the isthmus; chapter 2 discussed the reciprocal separation between marine biotas, while chapter 5 treats changes in terrestrial plant life. Because the fossil record is more difficult to follow for plants, chapter 5 focuses on changes during the past glacial cycle only. Together these three chapters demonstrate the profound biological consequences of the grand geological changes that Central America has experienced during the past 3 million years.

Five questions about Central America's role in the interchange that linked the land biotas of North and South America stand out:

1. What kind of evidence is used to identify the interchange?
2. When did it begin?
3. Which species were involved?
4. What has happened to the animals of the interchange today?
5. How did humans subsequently affect the faunas involved?

What Kind of Evidence Is Used to Identify the Interchange?

I want to concentrate on vertebrate animals, as these have left the most extensive terrestrial fossil record through time and space, and they contain the most complete information about the ecological conditions under which they lived. There are two large data sets that hold lessons about land animals in Central, North, and South America, one from modern samples of living animals, the other from fossils.

The most fundamental and accessible data are those recording current distributions of present kinds of animals. These modern data are compiled from many explorers and scientists who have made observations and collections since Columbus's discovery of the Americas. Generally these modern distributional data are very reliable. For example, records of the jaguar (*Panthera onca*) from both sightings and museum specimens are widely recorded.

Surprisingly, however, for some groups of living vertebrates distributional data remain incomplete. Some animal species are still undiscovered even as many others are becoming extinct. For example, within the past decade an entirely new genus and species of large animal was found in the Gran Chaco of Uruguay, Paraguay, and Argentina. It is a peccary (large piglike animals) with the scientific name *Catagonus wagneri*, and it is the size of a goat. By the time this animal was discovered by scientists, it had been hunted and displaced by land clearance almost to extinction. If modern scientific recognition of this large, wide-ranging animal was so nearly missed, one can only guess at how many smaller forms have gone, or will soon vanish, without any record. Current losses occur at an alarming rate wherever human populations are expanding. The largest losses are in equatorial rain forests because they hold the greatest diversity on land.

The second source of distributional data is that developed by paleontologists. The term *paleontology* means ancient (*paleos*) life (*onto*) study (*logo*). It combines the study of earth (geology) with the study of life-forms (biology). For many reasons, searching for clues of ancient life is far more difficult than searching for modern species. A major advantage, however, is that the fossil data usually carry with them a chronological dimension and include records of animals that became extinct long ago. Fossils thus produce a deep historical record of the past existence and distribution of plants and animals. Generally, marine records of mostly invertebrate species (see chapter 2) are more extensive than terrestrial records of mostly vertebrate species. Fortunately, Central, North, and South America have yielded excellent paleontological records of land animals from the past few million years. These records demonstrate the changing distributions of species before, during, and after the time the American continents were conjoined. Several key fossil vertebrate sites, especially in Central America, have been located (fig. 4–1).

In the highlands of Guatemala and Honduras, the core of an ancient continental mass, one can see magnificent geological vistas. For exam-

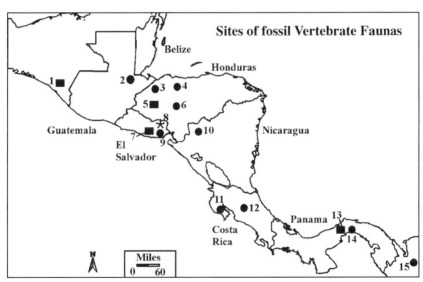

4–1. Locations of the principal sites in Central America of vertebrate fossils containing faunas involved in the Great American Biotic Interchange: (1) El Gramal, Mexico; (2) Patterson Site, Guatemala; (3) Yeroconte, Honduras; (4) Olancho, Honduras; (5) Gracias, Honduras; (6) Humuya, Honduras; (7) Corinto, El Salvador; (8) Arroyo del Sísmico, El Salvador; (9) Hormiguero, El Salvador; (10) León, Nicaragua; (11) Nacaome, Costa Rica; (12) Barrantes, Costa Rica; (13) La Cucaracha, Panama; (14) Coca, Panama; (15) Chivolo, Colombia. Asterisk = peak of interchange (Late Pliocene) with mingled North and South American vertebrates; squares = preinterchange (Miocene) with North American vertebrates; circles = postinterchange (Pleistocene) with mingled North and South American vertebrates.

ple, the view from El Celaque, the highest peak in Honduras, south-eastward to Santa Bárbara with its rich gold and silver mines soars across a deep, rectangular valley that leads northward from the ancient stronghold of Gracias. About 1536, Gracias was the scene of the outbreak of a revolution against Spain led by the indigenous leader Lempira, and it is now the capital of the remote Department of Intibucá. The mountains are ancient, but the intervening terrain contains vast volcanic flows about 15 million years old, forming the High Volcanic Plateau.

Within the basin surrounding Gracias lies an outstanding fossil record of land life in Honduras from about 10 to 8 million years ago, during the middle Miocene. During that time the basin had slowly filled with ashy sand and volcanic detritus as it was downfaulted and eroded. These gritty sediments accumulated rapidly enough to preserve an abundance of three-toed horses and long-jawed mastodonts. Repeated collecting campaigns by several museum teams have expanded the list to include rarer members of this fauna, including camels, rhinoceroses, peccaries, dwarf deer, and short-faced wild dogs.

Corinto, a similarly aged site in the foothills of El Salvador, confirms the presence of many of the same animals. Two of the unique Miocene fossil species from Corinto and Gracias, the short-faced (hyaenoid) dogs and the three-toed horses, were commemorated on postage stamps issued by El Salvador in the early 1980s. Farther north, near the Pacific edge of the Isthmus of Tehuantepec, is El Gramal, another site with similar Miocene dogs, camels, and horses. It provides useful evidence of biological continuity between southern Mexico and the High Volcanic Plateaus of Honduras and El Salvador.

In the course of studying the Panama Canal excavations, paleontologists discovered evidence in eastern Panama of more terrestrial vertebrate life during the middle of the Miocene epoch. The main site, La Cucaracha, has produced extinct genera of rodents, horses, rhinoceroses, and two extinct families of camel-like ungulates (hoofed mammals), one bearing horns on its head. Each of the Panama forms bears a close resemblance to animals of similar age known from Texas and Nebraska, demonstrating the North American affiliation of the Panama fauna. The most diagnostic fossils were referred, even down to the species level, to northern counterparts. Surprisingly, the degree of resemblance then was far greater than it would be in a similar comparison of living vertebrates today. The water barriers that divided Central America into a large archipelago during the middle Miocene (see fig. 1–19) evidently were not so wide and were not swept by such swift currents that they prevented occasional swimming or rafting between adjacent land areas. Furthermore, the Central American islands must have been rather uniformly congenial to a variety of land animals that spread easily and frequently enough to maintain close genetic continuity.

The continuity between terrestrial faunas of North and Central America contrasts strikingly with the separation between those of Panama and South America. There, a deep-water portal between the Atlantic (Caribbean) and the Pacific was still very wide and swift, serving as an absolute barrier to land life. In the upper Magdalena River of northern Colombia, the area known as La Venta produces a rich land mammal fauna of middle Miocene age. It contains no species (or even any families) of mammals in common with contemporaneous North or Central America. All species belong to the South American fauna, which remained totally isolated until the very late Miocene.

Then, about 8 million years ago, things began to change. By looking carefully at fossil collections in North and South America, one can detect three subtle hints of events to come. In the late Miocene two kinds of ground sloths appear in North and Central America, and at about the same time, one kind of raccoon (now extinct) appears in Argentina. Until that time, all ground sloths and their relatives were exclusively South American, and all raccoons and their relatives were exclusively North American. This evidence gives the first hint that the deep-water currents had weakened. But this was not a major break; no other examples of barrier crossings can be discovered until much later. Ground sloths and raccoons are durable swimmers and broadly adapted to living in a wide variety of environments. Raccoons are well adapted to swimming and feeding in streams and ocean margins, and it is not uncommon to see living tree sloths calmly swimming across Amazon rivers or Lake Gatún in Panama. Both groups also reached Cuba, Jamaica, and other islands of the Greater Antilles in the Miocene. Thus, it is probable that these early heralds of the interchange indicate the formation of an archipelago (see chapter 1) with widely spaced islands that allowed only the best swimmers to island-hop across to South America. The absence of other interchange species shows that there was nowhere near a complete land bridge.

No further evidence of an inter-American faunal connection or barrier crossing appears anywhere in the Americas until less than 3 million years ago (in the late Pliocene). Then vast numbers of new emigrants appear abundantly on both sides of the great barrier. It is overwhelmingly clear that a continuous land bridge arose, providing a continuous land route for a dozen families and for many dozens of genera to pass in both directions. This was the Great American Biotic Interchange. Before looking more closely at the many kinds of vertebrates that joined in the reciprocal northward and southward dispersals between the Americas, it is important to specify the exact timing of this remarkable event.

When Did the Interchange Begin?

The question of when the land bridge between the American continents was completed has been answered in various ways by different disciplines. Before comparing dates, therefore, one must consider carefully what is being dated. Jeremy Jackson and Luis D'Croz, studying the marine seaway, date strata in which marine organisms diverge and reflect shallowing and severing of interocean continuity and also upwelling of adjacent currents. Predictably, these disruptions of the formerly broad and deep portal between the Atlantic and Pacific must precede establishment of a complete isthmian land bridge across the former seaway. Logically, one would expect the earliest dates leading up to the land bridge to come from deep-water marine separation, then shallow-water settings, and finally from terrestrial connections between North and South America.

These expectations are borne out by the data from independent disciplines. For many years the marine separation dates, ranging between 3 and 4 million years ago, were widely quoted. The younger dates for establishment of the land bridge, about 2.5 million years ago, were derived from land mammal dates and were often thought to be too young because they were only about half as old as the oldest marine-derived dates. Taken together, these dates reflect the whole history of the closing seaway and the emerging land bridge.

The terrestrial dates for completion of the land bridge are based not only on fossil evidence from Central America, but also on a broad sample of dates wherever the waves of new immigrants appeared in North and South America. At present the most precise chronology dating immigrants that surely crossed the isthmian land bridge can be found in the western United States. In southern California, where the San Andreas fault passes through the Salton Sea basin and into the Gulf of California, active rifting provides a rich geological section of marine and terrestrial sediments. The Imperial Formation of late Pliocene age (about 3.5 to 2 million years ago) includes tropically adapted monk seals and many warm-water invertebrates that had continuity from the Atlantic through Panama's marine portal to the Pacific Ocean and then north to what is now southern California. Just above the Imperial Formation, however, in slightly later Pliocene terrestrial deposits, occur diverse immigrant land animals from South America, including armored tank–like glyptodonts, armadillos, more sloths (not closely related to the ground sloths that came earlier), capybaras (large aquatic rodents still living in the tropical lowlands), and porcupines (another group of South American rodents related to guinea pigs). Thus, the late Pliocene emplacement of the Central American land bridge is well documented in southern California

A similar suite of immigrants from South America is very precisely dated with volcanic ash dates and magnetic reversal chronology in southern Arizona in an area known as 111 Ranch. There, the best dates just below and just among the first evidence of immigrant animals from South America are about 2.4 million years ago.

In Central America, the best-dated section that gives evidence of new immigrants from South America occurs in the ancient lake beds near Arroyo del Sísmico in El Salvador. There, fine-grained ashy sediments of late Pliocene age delicately preserve many kinds of fossils, including such aquatic microorganisms as diatoms and ostracods (tiny bivalved crustaceans), and also plant impressions, fish remains, the whole skeleton of a rare mustache bat, and diverse large mammals, among them two previously undescribed kinds of ground sloths.

Similar dates for immigrant land animals occur in South American sites. The great sea cliffs southeast of Buenos Aires provide an excellent sequence of late Pliocene and early Pleistocene sediments full of land mammal fossils. There, the first evidence of groups derived from North America, including llamas, horses, sabercats, bears, peccaries, and field mice, appears in the very late Pliocene. The best dates fall rather close to 2.4 million years ago, just as they do in western United States and Central America.

Earlier evidence of terrestrial immigrants moving between the two American continents may be found. Until that happens, however, the sites just enumerated give an excellent approximation of the time the great interchange began. The time elapsed while several animal species spread through Panama northward to Arizona, in the absence of major barriers, is almost instantaneous in the perspective of geological time. The great speed of immigrant mammals is exemplified by rabbits, which, introduced into Australia for the sport of hunters, spread over the entire continent within a few decades. Although the search for earlier contact continues, an abundance of evidence throughout the New

World places the first rush of interchange animals through the isthmian area at about 2.4 million years ago.

Which Species Were Involved in the Interchange?

Once the land bridge was established, an extraordinary crossing of many terrestrial animals began in both directions. The most comprehensive fossil evidence comes from land mammals because they are generally abundant and their teeth, which fossilize particularly well, make possible rather precise identifications, usually to species level. A general accounting of families that are recorded on both sides of the land bridge in the late Pliocene and early Pleistocene is given in tables 4–1, 4–2, 4–3. Most of these groups are known to have extended their geographic ranges right through the tropics and into temperate latitudes of the opposite continent. Llamas, for example, which have Miocene and early Pliocene records only in North America, are found (suddenly) in the late Pliocene in the southern semidesert terrain of Patagonia. Equally remarkable is the widespread distribution of the ground sloth *Megalonyx*, which reached both coasts of North America and as far north as Alaska.

Can one infer from this roster of families that crossed through Panama what the land bridge was like ecologically at the time the bridge formed? Most groups were grazers (eating grasses and other coarse, low herbs) that lived in large herds and are best known in settings of open woodlands or grassland savannas. Horses and llamas are two of the most familiar examples from North America. From South America, toxodonts (large rhinoceroslike ungulates), diverse ground sloths, and glyptodonts (shelled relatives of armadillos with massive plant-grinding jaws and teeth) suggest similar ecological settings. Some herbivorous mammals of the interchange were browsers or mixed feeders, but a majority were grazers.

Not all interchange mammals were herbivores, however. Six families of carnivores entered South America via the land bridge with devastating effect on the native herbivore populations, which had absolutely no experience with efficient mammalian carnivores. Within a geological instant after the interchange began, a richly mingled fauna ranged widely from the Great Plains of North America to the Pampas of Argentina and was surely well adapted to temperate grassland savanna and to open woodland. Thus, both continents were faunally enriched by the late Pliocene interchange.

Central America hosted a particularly vital mixture of new immi-

Table 4–1. Mammal Families of the Great American Faunal Interchange

Legions of the *North*	
Scientific Name	Common Name
Soricidae	shrews
Leporidae	rabbits
Heteromyidae	pocket mice
Geomyidae	pocket gophers
Sciuridae	squirrels
Cricetidae	field mice
Felidae	cats
Mustelidae	skunks and otters
Canidae	foxes
Procyonidae	raccoons
Ursidae	bears
Gomphotheriidae	mastodonts
Tapiridae	tapirs
Equidae	horses
Tayassuidae	peccaries
Camelidae	llamas
Cervidae	deer

Table 4–2. Mammal Families of the Great American Faunal Interchange

Legions of the *South*	
Scientific Name	Common Name
Didelphidae	opossums
Dasypodidae	armadillos
Chlamytheriidae	giant armadillos
Glyptodontidae	"tanklike" edentates
Megalonychidae	bear-sized ground sloths
Mylodontidae	middle-sized ground sloths
Megatheriidae	elephant-sized ground sloths
Bradypodidae	three-toed tree sloths
Myrmecophagidae	anteaters
Callithricidae	marmosets
Cebidae	monkeys
Hydrochoeridae	capybaras (large aquatic rodents)
Erethizontidae	porcupines
Caviidae	guinea pigs
Agoutidae	pacas
Dasyproctidae	agoutis
Echimyidae	spiny rats
Toxodontidae	rhinoceroslike ungulates
Phororhachidae	giant predaceous birds

Table 4–3. Families Which Go Extinct in the Pleistocene of
Central America*

Chlamytheriidae
Glyptodontidae
Mylodontidae
Megatheriidae
Hydrochoeridae
Gomphotheriidae
Elephantidae
Equidae
Camelidae
Bovidae

*Some groups listed here survived in South America. In addition many genera (e.g.,
Smilodon) became extinct, but they are not listed if their family (e.g., Felidae) survived in
Central America.

grant animals extending their ranges in both directions. Scientists cannot specify what the crossroads were like ecologically in the late
Pliocene, but it is clear that the interchange landscape included a wide
range of both forested and unforested habitats, offering a broad ecological avenue to many kinds of land animals.

The many large herds of grazing and mixed-feeding herbivores
themselves had a strong ecological impact on Central American landscapes, as do the vast herds of ungulates in the tropical and subtropical
savannas of Africa today. Massive annual migrations, then as now, allowed diverse groups to alternate grazing on coarse fodder with mixed
feeding and browsing on forest margins during the most favorable seasons. Such high activity undoubtedly kept all but the densest rain forest more open than at present. Modern tropical ecologists point out that
the large fruits of palms and guanacaste trees require large herbivores to
crack and disperse them. Horses, reintroduced by the Spanish, are the
only competent seed-cracking herbivores at present. One can only
guess at the elaborate interactions that must have occurred between
large herds of herbivores and tropical American vegetation during the
Pliocene and Pleistocene.

Paleontological evidence of the interchange throughout North,
Central, and South America suggests that diverse species extended
their ranges rapidly and then stabilized in the early Pleistocene. Most
families that migrated had done so during the late Pliocene or early
Pleistocene. According to the North American record, the last straggler
was the opossum, which arrived in Florida in the mid Pleistocene, just
over 1 million years ago. Horses, camels, sloths, peccaries, toxodonts,

and many others had crossed reciprocally through the tropics and lived in both American continents by about 2 million years ago.

Recent animal distributions, as well as some late Pleistocene fossil evidence, suggest that there was a second, ecologically different phase of the interchange, one in which Central America played a decisive role. The second phase involved a different cast of characters, animals that were more specifically adapted to equatorial lowland rain forest environments. They did not extend their ranges into high temperate latitudes on either side of the equator. These animals moved mainly from the vast domain of the Amazon basin northward throughout the lowlands of Central America, reaching their northernmost limits along the Caribbean coast below the tropic of Cancer, in the general area of Veracruz, Mexico. This second stage of intertropical faunal movement is represented among birds by parrots, toucans, and guans (arboreal, tropical turkeylike birds), and among tropical butterflies by the magnificent iridescent *Morpho* and the colorful *Heliconius*. Tropical mammals that moved from Amazonia to Central America include tree sloths, large lowland rodents such as agoutis and pacas, and a myriad of monkeys.

The wealth of tropical biota that moved northward into Central America in this second wave of the interchange shifted the dominant faunal characteristics from North American temperate to South American tropical. And that is why the great nineteenth-century naturalists such as Charles Darwin, Alfred Wallace, and Joseph Hooker, in surveying the New World's rich terrestrial fauna and flora, linked the Central American tropical biota closely with the Amazon biota, placing them together in the *Neotropical Realm*. All subsequent biological surveys and studies of the American tropics have confirmed the validity of this Neotropical affiliation. Only geologists and paleontologists are aware of the earlier close linkage between North and Central American land life and its total separation by a deep-sea barrier from South American lifeforms.

Paleontologists can confirm this second intertropical phase of the interchange mainly by negative evidence. Some large grazing animals from North America that might have been expected to cross the isthmian land bridge into South America did not do so because they came too late (see below). These include pronghorn antelopes, mammoths, and the American bison. The earliest bison (also known as buffalo) entered North America from Asia in the late Pleistocene and spread widely not only throughout the grasslands of the midcontinent but also into open woodlands in the east and southeast. Bison herds also spread southward into Central America, following thornscrub and savanna habitats along the Pacific slopes of El Salvador, Honduras, and northern Nicaragua. Indeed, some remarkable footprints of bison and early humans are preserved on the surface of an ancient lava flow where the lava and the fleeing creatures entered the northern edge of Lake Nicaragua. However, bison were not present in Central America early enough in the Pleistocene to spread southward through the more open environments that may have prevailed then. Evidently they were too late: the mesic climate and massive rain forest prevented them from reaching southern Central America and crossing the land bridge. Such a conclusion coincides with Paul Colinvaux's evidence (see chapter 5) that in the later Pleistocene lowland rain forest throughout Panama presented a massive barrier to open-country animals.

Several biological lessons inhere in the subsequent history of groups that participated in the interchange. In general, the groups that spread southward into South America had a long, wide-ranging history not only in North America but also, before that, in Asia. They were in some sense already successful in spreading through a wide range of latitudes from one continent to another. And most of these groups from the north became very successful, by spreading widely and diversifying greatly after they had entered South America. Groups extending their ranges in the opposite direction (into Central and North America) had no other continental experience, having been isolated in South America by ocean barriers. The following examples will show that they had very limited success, undergoing virtually no diversification and only one or two wide distributions.

Animals That Spread Northward

Megalonychid Ground Sloths

These bearlike animals were one of six families of sloths that evolved in South America during its long isolation from other continents. Their cousins, the much smaller tree sloths, still survive hidden in the upper canopy of equatorial rain forests in Central and South America. Four distinct branches of ground sloths, including megatheriids such as *Eremotherium* (fig. 4–2), which were much larger than megalonychids, reached North America. But the megalonychids came first and spread widely. As stated above, some megalonychids were the first to cross narrow water barriers about 8 million years ago, anticipating by more than 5 million years the construction of a complete isthmian land bridge. Descendants of these sloths spread more widely throughout temperate North America than any other immigrant group from South America. In the late Pleistocene they occur as far north as Alaska and in almost every state in the contiguous United States. President Thomas Jefferson was intrigued by those that reached Kentucky and were discovered at Big Bone Lick when that was a frontier area. That large, late Pleistocene species was named *Megalonyx jeffersoni*. The largest megalonychid occurred much earlier in Central America. That rare sloth is *Meizonyx salvadorensis,* a primitive form with narrow, pointed canine teeth from a very late Pliocene site at Arroyo del Sísmico in El Salvador. Unfortunately, all of the ground sloths throughout the Americas became extinct at the end of the Ice Age. President Jefferson had hoped that Lewis and Clark would find some when they explored the western wilderness and specifically instructed them on that mission. Today there are few places remote enough to look for living examples of such large animals.

The Giant Anteater

One of the most distinctive families of large mammals that evolved in South America consists of three genera of anteaters. The biggest and most distinctive of these is the giant anteater, with the scientific name *Myrmecophaga tridactyla* and the equally appropriate Spanish popular handle, *oso hormiguero* (*oso* = bear, *hormiga* = ant) (fig. 4–3A). Nearly every feature of its anatomy and behavior is specialized for its unusual diet, which consists exclusively of termites from large terrestrial mounds that develop abundantly on fallen trees and tree roots in tropical American savannas. Among the giant anteater's peculiar adaptations are its powerful claws, short, heavy limbs, long hair, extremely long tongue, and tubular jaws with no teeth. Early fossil records of almost modern myrmecophagids come from Brazil, Uruguay, and Colombia about 15 million years ago. Of course, none are known in North or Central America until after the land bridge allowed their northward progress. Then clear-cut evidence shows that *Myrmecophaga* shuffled and ate its way clear through the American tropics in time to be fossilized in a 1.5-million-year-old site known as El Golfo in the northwestern corner of the state of Sonora, Mexico. This fossil antbear reached about 3000 miles north of its nearest living relatives, which now occur in the tropics of eastern Guatemala and southern Belize. The present climate in this part of Sonora is too arid and too cold in the winter to support the tropical forests that sustain an abundance of large termite nests, which is the only food that can sustain a population of

Eremotherium

4–2. The giant ground sloth *Eremotherium* scaled to a human 2 meters tall. Sloths are migrants from South America.

giant anteaters. The giant anteater ranges southward into northern Argentina.

Armadillos

One of the South American groups that repeatedly extended its range northward into Central and North America was the armadillo family. Perhaps their shells protected them from unaccustomed predators, and perhaps their broad range of acceptable diets, including insects and other small animals, carrion, tubers, fruits, and other plant parts, helped

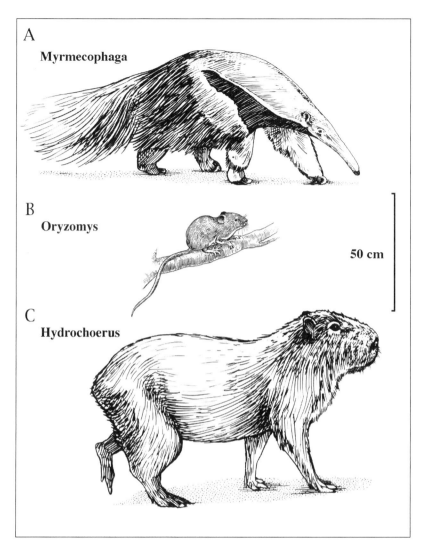

4–3. (A) The giant anteater *Myrmecophaga,* a southern migrant; (B) *Oryzomys,* a rice rat, one of the vast variety of the mice family that proliferated in South America after migrating from the north; (C) *Hydrochoerus,* the giant capybara rodent, evolved from migrants from the north.

them survive in all but the most arid environments. The only modern species in temperate North America (north to Oklahoma and Kansas) is *Dasypus novemcinctus* (the nine-banded armadillo), but several other extinct relatives reached equally far north in the late Pliocene and Pleistocene. In Central America the nine-banded armadillo is joined by the naked-tailed armadillo, *Cabassous centralis.*

One of the largest and most impressive of the extinct relatives of armadillos was *Holmesina,* the tanklike form with a shell nearly four feet in diameter (fig. 4–4B). Glyptodonts specialized in eating plants and plant roots and had much deeper jaws and more powerful cheek muscles than any living armadillo. Some specimens show a deep stab wound into their skulls, indicating that they were hunted and killed by saber-toothed tigers. In spite of this evidence that some individuals were mortally wounded, the glyptodonts were among the first species to cross the land bridge and establish themselves abundantly throughout North and Central America. On the other hand, their importance may be somewhat exaggerated because their massive shell elements are easily preserved, discovered, and identified.

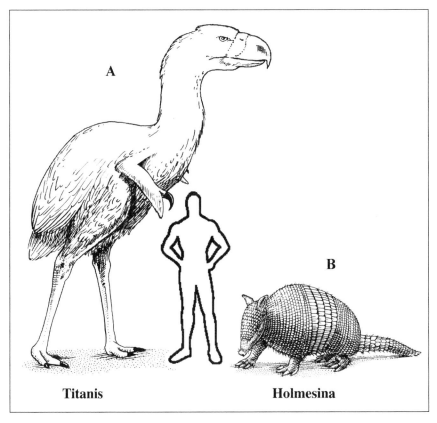

Titanis   **Holmesina**

4–4. (A) *Titanis,* the spectacular, giant, carnivorous bird scaled to a 2-meter human. A migrant from the south, they are found fossil as far as Florida in North America; (B) *Holmesina,* one of the giant armadillos, is also a southern emigrant.

Toxodonts

Most of the large herbivorous animals that had evolved in South America vanished before the interchange or were vanquished by competition from such northern immigrants as horses, camels, and tapirs. One notable exception, however, was the toxodonts, a kind of rhinoceroslike group of very large herbivores (figure 4–5A). The name *toxodont* refers to their curved upper teeth, which formed a powerful battery of tall grinders, well equipped to masticate large volumes of leafy material. These large herding herbivores actually extended their ranges northward against the tide, as it were, of northern ungulates moving south. The species *Mixotoxodon larensis* has been found in late Pleistocene deposits in every Central American country, usually in association with mastodons, giant ground sloths, and horses.

Giant Birds

A most unexpected South American participant in the interchange is a gigantic predaceous bird known as *Titanis walleri* (see fig. 4–4A). When the first fossilized toe and claw bones were discovered in the bottom of Florida's Santa Fe River they were thought briefly to belong to a dinosaur. Indeed, their powerful hind feet for running and tearing prey and their large size (more than three meters tall) closely resemble those of some dinosaurs. Their presence in late Pliocene sediments along with sloths and glyptodonts, however, indicated that they must be compared with large birds, and when scholars turned their attention to South America possibilities they quickly discovered that *Titanis* was a new genus belonging to the extinct family Phorusrhachidae from Brazil and Argentina. In recent years many more parts of this remarkable predaceous bird have been discovered in Florida, including its meat-cleaver beak. Evidently members of this group followed some of their traditional prey (such as glyptodonts) northward through the savanna corridors of tropical South America and Central America for a distance of more than 10,000 kilometers.

Animals That Spread Southward

The Cat Family

Before the land bridge formed, when South America was still an isolated continent, it had no efficient mammalian carnivores. The giant birds, discussed above, were the major large predators in open-country

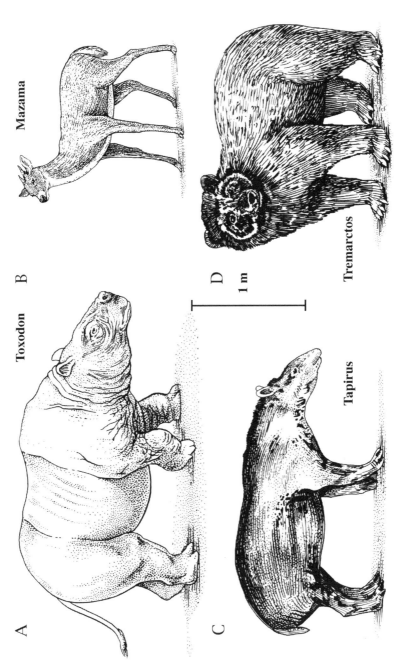

4-5. (A) *Toxodon*, a large herbivorous animal from South America that evolved in South America after migrating from the north in the interchange; (C) *Tapirus*, a well-known herbivore in Central America and the Amazon that evolved from North American stock; (D) *Tremarctos*, the bespectacled bear distinctive of the American tropics and descended from northern emigrants.

habitats, while crocodiles were active in aquatic settings. It is not surprising, therefore, that when efficient, warm-blooded mammalian predators entered South America they became widespread and successful in virtually all habitats.

This discussion focuses on the cat family, but four other carnivore families, bears, dogs, weasels, and raccoons, also extended their ranges through Central America, crossed the land bridge, and reached South America. All of these carnivore families except the bears include a considerable diversity of living genera and species in the American tropics, probably representing multiple dispersals. The bears are represented in the interchange by the spectacled bear, which still lives in tropical south America.

Both large and small members of the cat family (Felidae) prospered in the new southern continent. A surprising number of small species coexist in South America, including margays, ocelots, jaguarundis, little spotted cats, and also *Felis guigna, geoffroyi,* and *colocolo.* Many are grouped under the popular name *tigrillos.* Most, if not all, of these small felids had already originated in North America before the land bridge was formed. It is likely that margays, ocelots, and little spotted cats, which specialize in tropical arboreal habitats, had already developed in Central America. They were thus ideally positioned to spread southward when the land bridge formed.

The two largest American cats are the puma and the jaguar (fig. 4–6). There is also the rare Andean cat *Felis jacobita.* The jaguar (*Pan-*

4–6. *Panthera,* or jaguar, the largest of the tropical American cats. Cats diversified into many species in Central and South America after the interchange.

thera onca) is the largest American cat and much admired for its magnificent glossy, spotted coat. Its coat is reminiscent of a leopard's in the Old World and is similarly adapted to camouflage the large cat not only during the day in shadowy jungle settings, but also at night when it does most of its stalking. As fossils, jaguars are known in Florida, where the largest individual skulls are known, and from the time of the interchange into historic times they ranged from southwestern United States into northern Argentina. They are now restricted to large areas of fairly dense vegetation.

The puma (*Puma concolor*), on the other hand, ranges through virtually all habitats in the Americas and extends from extreme northern to extreme southern latitudes. Before the land bridge formed, both panthers and jaguars lived in Central America. Less than 3 million years ago, the land bridge offered them major new opportunities in South America. Jaguars extended their ranges through the Amazon basin into Argentina and Uruguay, while panthers covered the whole new continent.

South American Deer

One of the most successful groups of large mammals to go south is the deer family. Deer originated in the north temperate latitudes of Eurasia some 20 million years ago. They were browsers and mixed feeders that lived on forest margins. About 5 million years ago distant relatives of the white-tailed deer reached the Bering Strait, while it was still forested, and entered North America. During the Pliocene they reached Central America and there diversified. Until then, deer had never reached tropical latitudes. Important new branches of the deer family that originated in Central America were the red and brown brocket deer (genus *Mazama*) (fig. 4–5B), characterized by small body size and simple spike antlers. Another new group of North American deer developed moderate body size and short, powerful legs as an adaption to mountain dwelling. They included an extinct genus from western North America known as *Navahoceros* and the living Andean deer (*taruca* and *huemul*) with the scientific name *Hippocamelus*. The name reflects the uncertainty of its original describer as to whether it was a horse or a camel; imagine how surprised that early scientist would be to learn that he had named a deer living in the southern hemisphere. When lowland rain forest closed in the isthmian corridor, the former continuity between the short-legged deer *Hippocamelus* to the south and *Navahoceros* to the north was broken.

After the interchange still other new kinds of deer arose in South

America. The surviving examples range from the diminutive *Pudu* of Andean forests through *Ozotoceros,* the pampas deer (similar in size and general appearance to the white-tailed deer), to the immense marsh deer (*Blastocerus dichotomus*) of moist tropical and subtropical habitats. And these new kinds of native deer were accompanied by immigrant groups like the white-tailed deer and the brockets from Central America. Thus, thanks to the interchange, South America changed from a continent with no deer to the continent with the most kinds of deer.

An Explosion of Mice

Even more successful than the deer family were the mice of the family Cricetidae. Before the interchange there were none in South America; today there are some sixty species in at least forty genera (see fig. 4–3B). Unquestionably these mice crossed the isthmian land bridge as part of the interchange, but their subsequent success is so astonishing that many biologists have questioned the timing of these events. How, they ask, could these mice have developed so many branches? Some are terrestrial, some arboreal, some pastoral (grazing), some sylvan (browsing), and one group (subfamily Ichthyominae) is uniquely adapted to catching fish in Andean streams. Paleontologists, constrained by the age of the land bridge, suggest that these mice experienced much of their diversification in Central America before the land bridge formed. At least six different branches (subfamilies) of the family Cricetidae are known as fossils in the late Miocene and Pliocene of subtropical North America, and many of them resemble their living South American descendants. This means that the time available for attaining the immense diversity of mice now seen in South America was about 9 million years (since the late Miocene), more than three times longer than the time available since the land bridge formed. Many cricetid adaptations to tropical conditions were probably developed in Central America, a fortuitous staging place for their subsequent peregrinations in South America.

What Has Happened to the Animals of the Interchange Today?

I have examined some examples of the many groups that crossed in both directions when the interchange took place. Most groups that reached South America were successful in the sense that they spread widely and diversified, some considerably. Fully half of the land mammal genera living in South America today came from ancestors that crossed the land bridge from Central America less than 3 million years ago, and the same is true of the tropical fauna of Central America.

On the other hand, the temperate North American fauna is decidedly lacking in any strong carryover from South America. Only three mammal species from the interchange remain north of the tropics, namely, the opossum *(Didelphis virginiana),* the nine-banded armadillo *(Dasypus novemcinctus),* and the porcupine *(Erethizon dorsatum).*

In many respects, however, the success of South America mammals in temperate North America has been concealed today by a vast cataclysm (see below) that caused the extinction of most large mammals at the end of the Ice Age. This loss is much greater with respect to South American mammals than to North American forms. All of the great ground sloths and their shelled cousins the giant armadillos and glyptodonts suddenly vanished in North America, as they did in South America. Likewise, the giant amphibious rodents known as capybaras (see fig. 4–3C) disappeared from North and Central America, where they had been abundant. The great toxodonts from South America had become well established in Central America but vanished along with all of South America's native ungulate stocks at the end of the Pleistocene.

Several families of North American ungulates were also devastated by late Pleistocene extinctions, but there were nevertheless numerous survivors, many of them in the American tropics. All the species of horses and of mastodonts in North and South America were lost. Llamas, peccaries, and tapirs and jaguars were exterminated in temperate North America, although they survived in tropical America and temperate South America.

What caused these late Pleistocene extinctions? They had little effect on smaller-sized mammal groups. The main impact was on large herbivores. Most carnivores survived. For example, among the cat family all but one kind still survive. One might have predicted extinction for pumas and jaguars because so many large game animals that were their prey disappeared late in the Ice Age. This implies that they may have relied more on deer and peccaries than on the many extinct kinds of large herbivores, such as horses and mastodonts. The one genus of large cats that became extinct were the sabercats (genus *Smilodon*). Their demise was surely predictable. The South American species had become extremely large during the Pleistocene. Such sabercats were bound to become extinct throughout the Americas when most of their large prey species died out.

American paleontologists and archaeologists have proposed two hypothetical causes of the mass extinctions of large mammals at the end of the Pleistocene. One hypothesis points to sweeping climatic changes that took place at the end of the last glacial epoch. The other implicates the hunting peoples who entered the New World and spread rapidly throughout the Americas. Both climatic events and the spread of humans correlate well with the time of mass extinctions. Dozens of kinds of large mammals, including ground sloths, horses, camels, mammoths, and mastodonts, disappeared almost at the same instant (geologically speaking), about 11,000 years ago. These extinctions may well have resulted from the combination of climatic shifts and human impacts.

The Role of Early Humans

Direct evidence of late Pleistocene animals being killed and butchered by human hunting is extremely rare in the New World. Most of the early records of Paleoindians throughout the Americas consist of lithic elements, predominantly fluted points like the classic Clovis and Folsom points (see chapter 6). Although these points were used in most instances to spear food, the lithics by themselves offer very little evidence of what species were hunted and how they were utilized.

Only a few dozen sites throughout the Americas provide ancillary evidence of the skeletal remains that were associated with early human inhabitants before the great extinctions about 11,000 years ago. And among these sites most direct evidence of animals that were hunted and butchered consists of elephantlike proboscideans. Different proboscideans were available in different regions of the New World, and it is now evident that human hunting bands were able to adapt to each new species they encountered as they spread through the New World. The first humans to cross the Bering land bridge from Asia relied heavily on the woolly mammoth (*Mammuthus primigenius*) not only for food but also for much of their economy (including fuel, clothing, and shelter). When they reached approximately 50 degrees north latitude (near the present United States/Canada border), they encountered the Colombian mammoth (*Mammuthus columbi*) and despite its different appearance, behavior, and habitat preferences, this species was heavily hunted. It is the prey species most often associated with classic Clovis sites in temperate North America. In the eastern United States from Michigan to Florida, the American mastodon (*Mammut americanum*) was also frequently utilized by Paleoindians. In Central America and

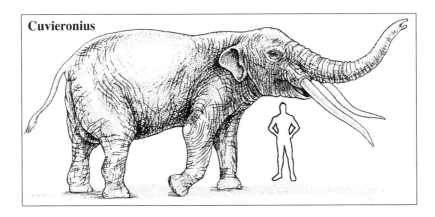

4–7. *Cuvieronius,* scaled to a 2-meter human, belongs to a family of Mastodons called Gomphotheres that evolved from northern ancestors in the American tropics.

South America, however, a distinct family of proboscidea (Gomphotheriidae), most frequently Cuvier's gomphothere (*Cuvieronius tropicus*), formed the primary large-animal quarry for the earliest humans (fig. 4–7). A few sites scattered through the Americas yield more detailed evidence of how these various proboscidean species were butchered and utilized by the first Americans. The tusks were modified to elaborate ivory tools, including pointed foreshafts that were evidently attached to the fluted flint points and to the long spears or javelins that were used to hunt other proboscidea. Bones were often made into such tools as digging hoes, shaft straighteners, and hide burnishers. Much of this bone and ivory technology can be traced back to late Paleolithic traditions of Eurasia.

It is difficult to extend the argument for human hunting of the late Pleistocene megafauna beyond the proboscidea. Very rare evidence indicates occasional use of horses. Most of the other evidence implicates heavy hunting of buffalo in North America and llama in South America, but these are species that survive to the present. Thus the hypothesis that dozens of large-animal species were hunted rapidly to extinction during the late Pleistocene is very poorly supported by direct evidence.

On the other hand, the absence of evidence does not constitute evidence that human hunting did not occur. Perhaps most of the remains of horses and other large mammals were largely consumed, leaving limited evidence. Much more work is needed in very late Pleistocene sites that preserve large animal bones. It is certainly possible that the earliest Americans did heavily hunt the megafauna and contribute substantially to the extinction of dozens of species of megafauna. This view becomes difficult to deny when one imagines these early bands employing clever hunting strategies: blinds, decoys, trained dogs, and fire drives.

The American tropics somehow prevented the extinction of a number of megafaunal species that became extinct at higher latitudes both north and south. Many species of great ground sloths, for example, vanished throughout both continents, yet two genera and several species of tree sloths continue their sublime existence in the canopy, oblivious of the terrible fate of their many cousins. More impressive are the world's largest rodents, the capybaras, pursuing successful careers in wet tropical settings. Likewise, tapirs (see fig. 4–5C), spectacled bears (see fig. 4–5D), jaguars, and several kinds of llamas and peccaries represent a substantial variety of large mammals that are cited as human-caused extinctions in North America but seem to continue quite unmolested in tropical latitudes. Because the same kinds of human hunters who traversed temperate areas also traversed the tropics in the late Pleistocene, these examples support the climatic hypothesis. Evidently, late Pleistocene extinctions acted more severely at temperate latitudes, where the probable effects were extremes of rapid climatic change.

To summarize, one cannot understand Central America today without knowing its extraordinary prehistory and geologic history. Neither can one comprehend the natural history of South America without realizing how strikingly it was affected by the land bridge that connected it with the rest of tropical America and beyond. The resulting interchange about 2.4 million years ago saw an immense reciprocal rush of new groups moving through the isthmian link to occupy new terrain in another continent. Many families were involved, about half of them inhabitants of relatively open country (savanna) landscapes. And many groups more typical of temperate latitudes moved all the way through the American tropics.

In the second half of the Pleistocene, starting about 800,000 years ago, the isthmian landscape became more fully occupied by lowland rain forest. Late-arriving grazing animals such as buffalo and mammoth could not migrate south of the seasonally arid scrub forest along the Pacific slopes of Nicaragua. Instead, the land bridge supported a strong northward surge of tropical (Amazonian) land life that spread abundantly through Central America and north, especially along the moist Caribbean coast to about the tropic of Cancer in central Mexico. This still marks the northern limit of the Neotropical Biogeographic Realm.

About three dozen species of large mammals involved in the interchange suddenly went extinct about 11,000 years ago. The extinction episode took place more severely in temperate North America and temperate South America than in the tropics of Central America and northern South America, which hints at rapid climatic change at the end of the Pleistocene as the primary cause. On the other hand, another major impact on large mammals at that time was the rapid spread of hunting human tribes from Asia into the Americas. There is substantial direct evidence of their effect on such proboscidea as mammoths, mastodonts, and gomphotheres. It is much less clear how (or whether) they impacted other possible game animals that became extinct.

More than half of the present land mammals of South America came from North and Central America by way of the land bridge. That is a startling fact to many biologists, who would not have thought that such extensive change could take place in less than 3 million years throughout such a vast and varied continent. The fossil record makes it quite clear that these revolutionary events did take place on this rapid timescale. The vast opportunities created when new groups are suddenly able to enter new environments produce extraordinary evolutionary change and diversification. The interchange provides unique insights into those dramatic evolutionary processes that cannot be studied when species live under more stable conditions.

CHAPTER FIVE

The History of Forests on the Isthmus from the Ice Age to the Present

PAUL COLINVAUX

Over the past 2 million years the climate of the earth has changed repeatedly in sympathy with the advance and retreat of the great ice sheets of the north. Central America never experienced extensive ice coverage, except on the tops of the very highest mountains, for example, the Altos de Cuchumatanes in Guatemala and the Talamancas in Costa Rica. But so great a convulsion as an ice age must have been felt even far away from the ice sheets themselves.

The coastlines certainly changed as the level of the oceans fell. Subtract from the oceans water volumes big enough to cover Canada and Scandinavia with ice a mile or so thick and the sea shrinks by more than 100 meters. Geologists confirm that sea level dropped by finding the old shorelines under the modern oceans.

So the isthmus used to be broader, running out to what is now the fifty-fathom line, and there was more lowland. But what were those ancient lowlands like before the sea submerged them? To find out, scientists look for traces of ancient vegetation in the sediments of ancient lakes. These sediments may span a variety of time periods, but the most complete sequences record the history since the last major glaciation about 19,000 years ago. This chapter, then, focuses on the changes that took place during this last half of the glacial cycle. I want to look first at the way in which modern tropical pollen is identified and distributed and then at the changes in pollen composition through time as recorded in various lake cores.

Mud settles to the bottom of a lake every year, so that a lake 20,000 years old has mud that can be imagined as 20,000 fine layers recording all the years of its existence. The settling mud takes to the bottom traces of the world about it, in its chemistry, its minerals, the microscopic remnants of life, and even its past magnetic field. The mud of ancient lakes is like a history book, its pages being the thin layers of mud. All that is needed by the paleohistorian is a drill to collect a column of mud from underwater and the ability to read the writing on those muddy pages.

The story of vegetation and climate is most clearly written in the pollen grains of plants. Over any plant community in the flowering season hangs an invisible cloud of pollen, undetected save by sufferers of hay fever, that is slowly settling as a pollen rain. Pollen grains that fall by the millions on the surface of a lake die, become waterlogged, and sink to the bottom. The contents of the pollen grains quickly rot, but the outer coat is made of one of the most resistant substances made by living systems. In the oxygen-starved atmosphere of lake mud, the husks of pollen last unaltered for thousands of years: millions if the lake should last that long.

The pollen of many kinds of plants can be distinguished with an ordinary light microscope. The analyst must first learn what the pollen grains of all the plants of a region look like, extracting pollen from plants in an herbarium collection and classifying their pollen type by type.

Lake mud might well have pollen grains in concentrations of 100,000 per cubic centimeter. After suitable treatment, a set of microscope slides is produced, each with several thousand pollen grains on it, which are then tallied by plant type until 200 to 300 grains have been counted.

Percentages of the different kinds of pollen have an obvious, but not direct, relation to the percentage composition of trees in the parent forest. Allowance must be made for the fact that some trees produce more pollen than others, but once that is done a remarkable reconstruction of the original vegetation can result. All that is needed for the history to be written is a series of radiocarbon dates to put a timescale against the pollen changes.

Pollen analysis was developed for use in temperate latitudes, where nearly all the pollen floating in the air comes from wind-pollinated plants, grasses, and, most important, trees. The trees of deciduous forests, like those of eastern North America, are nearly all pollinated by wind, and they produce huge amounts of pollen. A car top can turn yellow with pine pollen after being parked for a single day in

North Carolina. Because the species list of trees in the temperate forest is small, its pollen history can be read quickly as the changing mixtures of perhaps a dozen or so genera: oak, birch, elm, maple, hazel, pine, spruce, and the like.

Wonderfully detailed histories of the forests of eastern North America and Europe have been written in this way, using dozens or hundreds of lakes. These regions were near or actually under the glaciers, and their forest histories are ones of repeated invasions by one tree after another, for the entire 10,000 years of Holocene time since the ice melted. In these temperate lands near the ice the landscape changes were huge; none of the ice age vegetation was at all like that of the present.

Pollen can also be used to track the vegetational changes in Central America since the Ice Age, even though glaciers were not present there. For many years it was assumed that this would be very difficult. First, in tropical forests wind pollination is rare, as the more important plants are pollinated by insects, birds, even bats. The expectation was that these plants would not throw much pollen into the air. Second, tropical vegetation is so diverse that the task of learning all possible pollen types is immense. Last, finding ancient lakes in the tropics is difficult because lakes are abundant only in terrains whose drainage has been interrupted by the gouging and moraine-dumping action of extensive glaciers.

Researchers now know, however, that lake mud of the Darién and Amazon forests contains as much or more pollen than mud from Massachusetts or France. Trapping of pollen from living trees as it falls in the forest of Barro Colorado Island in Panama and in the Amazon has shown that the pollen fallout is indeed as massive as the lake mud suggests. This phenomenon remains a mystery, although perhaps forest trees are in such competition for pollinating insect carriers that they make grossly excessive amounts of pollen. Whatever the cause, this massive pollen production has meant that pollen analysis in the tropics has become an unexpectedly powerful tool in the reconstructing of past climates and vegetational history.

Four or five teams of pollen analysts have been working in Central America for some years now, an effort that has yielded histories of various lengths from the Petén, Belize, Costa Rica, the Pacific side of central Panama, and the Darién. The sites are as yet too few to permit the drawing of an Ice Age map of the isthmus or the compiling of a complete record of the changes that have occurred since. But answers to some of the bigger questions have become clear. The whole isthmus was colder, the modern forested coast of Petén and Belize was formerly arid, and the lowland forests of Panama persisted throughout the Ice Age, though much altered.

When a lake suitable for testing has been found in the isthmus, three rubber boats are deployed (all of which must be light enough to be backpacked into remote areas if necessary). Two of the boats are lashed together under a small wooden platform that serves as a drilling raft, and the third is used as a tender and tugboat. The raft is secured in the middle of the lake by three anchors on very long ropes, which have been dropped by the tender. A pipe is then lowered in sections to the bottom of the lake as a casing within which drilling can proceed. The casing allows the drill to be withdrawn and then replaced in the same hole repeatedly as the core is raised in sections. Drilling is a simple push-pull operation, with no rotation involved: threaded aluminum tubes, each a meter long and about 5 centimeters in diameter, are screwed together end to end and pushed or hammered into the mud. Cores up to 20 meters long have been obtained in water 40 meters deep.

How Did Climate Change in Central America during the Ice Age?

The first lake cores used to study the ancient vegetation of Central America were taken in the Petén of Guatemala. The pollen history there shows that before 10,000 years ago the forest had disappeared; the same has since been shown for Belize. What is now forested land on the Caribbean coast of northern Central America held only dry open vegetation during at least the last portion of the Ice Age.

Lake Valencia in northern Venezuela has yielded a similar history. What is now a big lake 40 meters deep was shown by an array of sediment analyses to have been a swamp in glacial times. The Petén and Lake Valencia results, coming from two points on the Caribbean coast that span the Central American isthmus, seem to suggest that the isthmus was significantly drier during glacial times and thus that tropical forests might have been absent or drastically reduced. In addition, as discussed in chapter 4, many of the animals that crossed the isthmus land bridge 2.4 million years ago were grazing animals—animals like ground sloths, giant armadillos, horses, and camels—that presumably required the savannalike conditions that increased aridity would produce.

Furthermore, this scenario appears to agree with the evidence from Africa, where increased climatic aridity in glacial times can be inferred from changed levels of water in almost any lake of great antiquity, especially those in East Africa. East Africa is, of course, a dry place at present, with severe dry seasons bounded by annual monsoonal rains.

It is entirely reasonable to suppose that the local climate was arid if monsoons were reduced during the Ice Age.

However, what was true of one tropical continent may not have been true of another. It appears that since the last glaciation 19,000 years ago, much of Central America has been more or less continuously covered with forests, although their composition and distribution were often changing. For example, important evidence for forest cover comes from lowland Panama, first from Lake La Yeguada, then from sediments of an ancient lake that now lies under the village of El Valle.

So far La Yeguada is the only permanent ancient lake found in Panama. About 17 meters of core were obtained, from which radiocarbon dates demonstrated that the past 14,000 years of history were represented (fig. 5–1). The period from 14,000 to 10,000 years ago was the time of transition from the full glacial period that ended 14,000 years ago to the final extinction of the glaciers 4000 years later. During this time, humans were on the move in the Americas, sweeping down from the Bering Strait, crossing the great plains, and reaching into the Central American isthmus. Lake La Yeguada sediments fortunately span this critical interval.

The first thing the La Yeguada sediments tell about this critical interval is that the land roundabout was never a semiarid savanna like the Petén. The pollen throughout is forest pollen, of changing composition certainly but never suggesting an arid landscape. Thus, the Pacific coast of Panama was probably moist enough to sustain forest through the time of the first human settlement.

But the La Yeguada cores provide evidence for moderately reduced precipitation all the same. This comes from the physical evidence of the lake sediments. La Yeguada is a closed basin lake and has never had a natural outlet, although plenty of small streams run into it; the lake level therefore changes with rainfall. The local climate is strongly seasonal, with 3800 millimeters of precipitation, 90 percent of which falls between May and November. In the absence of an outlet to set water level by spillover, the level of the lake should certainly change with the seasons. The sediments of the past 10,000 years are almost featureless lake mud, the typical well-stirred mixture of some clay from the banks with organic debris from the plankton and other lake life that we call *gyttja* (a word coined by pioneering Swedish lake scientists early this century). The thick sequence of deposits before 10,400 B.P. (before the present), however, is banded pink and green. On x-radiographs the pink bands show up whiter and the green bands blacker, evidence that pink bands have a higher inclusion of minerals that are opaque to x-rays.

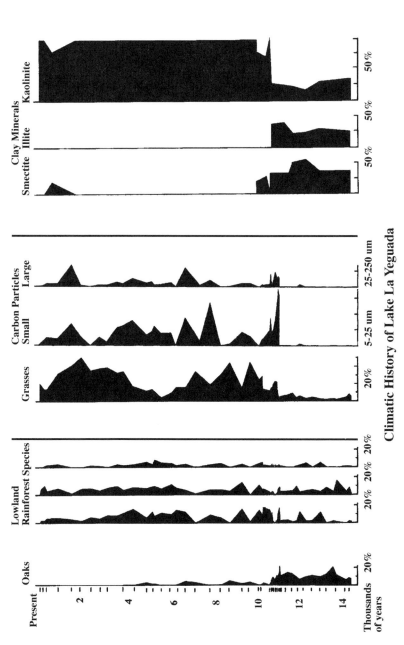

Climatic History of Lake La Yeguada

5-1. A pollen and clay mineral diagram of sediments cored in Lake La Yeguada (locality is indicated in figure 3–4). The diagram records the vegetational changes during the past 14,000 years. The horizontal axis shows the percentage of the total pollen of each type of tree, shrub, grass, or clay mineral. The vertical axis records thousands of years as calibrated by numerous radiocarbon dates. Particulate carbon is measured as 1000 grains per square centimeter. The diagram clearly shows that after 10,000 years, as postglacial warming occurred, oaks diminish and disappear and the clay minerals smectite and illite give way to kaolinite. Lowland tropical rain forest species are present throughout. As man appears some 11,000 years ago, burning generates substantial particulate carbon and grasses spread. From M. Bush, D. Piperno, P. Colinvaux, L. Krissek, P. de Oliveira, and M. Miller, "A 14,300-Year Paleoecological Profile of a Lowland Tropical Lake in Panama," *Ecological Monographs* 62 (1992): 251–75.

Most of the pink and green couplets are fine: 20 or more per centimeter but a few exceptional ones are thicker, which allows analysis of pink and green separately.

The green bands have significant amounts of chlorophyll pigments from algae, suggesting water of high fertility. Also, the green bands hold a dense array of the silica skeletons of microscopic plants called diatoms that are usually found in fertile waters. The pink, mineral-rich bands get their color from clay minerals characteristic of red tropical soils. Clay minerals are chemical products of weathering, particularly of the mineral feldspar. They have definite structures, microscopically platelike, with sheets of oxygen atoms stuck to sheets of iron or aluminum. A beam of x-rays impinging on a thin sample of a clay mineral is scattered like light through a prism, and the pattern of scatter is telltale evidence of the structure of the clay. The pink layers were rich in the clay mineral illite, but samples from the gyttja of the past 10,000 years have no illite, being mostly kaolinite.

Illite forms on a dry, exposed land surface, kaolinite forms in the wet. The alternating banding thus suggests that the La Yeguada watershed was more exposed to drying in the late-glacial episode of banding than it has been in postglacial time. Each pink-green couplet is now nicely explained as the manifestation of a dry-wet cycle. In the dry time, clays collect by synthesis on the soils of the watershed, with a notable component of illite, but also with nutrients like potassium or phosphorus stuck to them. When the rains finally come, some of the clays are washed into the lake to settle as a fine pink band. But the clays take nutrients with them, fertilizing the water. A bloom of diatoms and other planktonic life follows, depositing a green band of gyttja.

Thus, the pink and green banding in late-glacial time suggests extended dry seasons between the rains. The data are quite consistent with global climate models requiring a modest reduction in monsoonal rains, which is nicely reflected in the pink clay bands. But it was too modest a reduction of rains to abolish forest from the Panamanian lowlands.

The relatively minor loss of rains at La Yeguada is further illustrated by the fact that the lake level actually rose well above modern levels by 12,000 B.P., even though deposition of the pink bands had another two thousand years to go. So the La Yeguada sediment record shows that dry seasons were more prolonged 14,000 years ago, rains increased until 12,000 B.P., though prolonged dry seasons continued until 10,000 B.P. The wettest time of all was from 10,000 to 8000 years ago, when lake level was high and dry seasons no longer than those of the recent past. The climate over the past 8000 years has been more or less as we have known it in historic times. At no time in this history is there evidence that failure of the rains was sufficient to remove forest from the Panama lowlands.

A far longer record is given by the lake sediments lying under the village of El Valle, Panama, closer to the Pacific coast at just 500 meters elevation. Here, 55 meters of core were recovered (fig. 5–2). Radiocarbon dating has an effective range of about 30,000 years, and this limit is reached at about the tenth meter of the El Valle core. The bottom 45 meters of sediment, therefore, are older than this. The youngest date at the top of the column is 8000 B.P., recording when the old crater lake was drained, presumably by cracking of the volcanic rim of the crater. Thus, the record overlaps with the shorter history from La Yeguada but goes far back into the glacial period.

The El Valle sediments show that the lake never dried up through all the changes of the glacial cycle, indicating that the climate of the Panamanian coastal plain was never significantly arid for perhaps 120,000 years. The combined plant records from La Yeguada and El Valle strongly suggest, however, that 10,000 years ago the climate was notably colder. The principal evidence is that plants now confined to the Chiriquí Highlands then grew close to both lakes. Oak pollen percentages were high 14,000 years ago and earlier, leaving little doubt that oak trees grew in the La Yeguada basin and in the El Valle crater. In addition to the oak (*Quercus*) pollen there is pollen of such genera as *Cyathea, Alternanthera, Chenopodiaceae, Sapotacea, Thalictrum, Symplocus, Ranunculus, Valeriana, Gunnera*, and *Rumex*. Many of these plant genera have species that live in lowlands, but they are never found together except in tropical montane vegetation. The simultaneous presence of pollen of them all strongly suggests that many montane plants lived at low elevations in those days. And this montane pollen signature is strongly amplified by the data from phytolith analysis.

Phytoliths, literally "plant stones," are the silica deposits that many plants lay down inside their cells. The silica deposits in grass leaves are the best known, and they make necessary special teeth in grazing animals, like the continually growing molars of cows. But other plants deposit silica too, perhaps as a way of getting rid of it. The beauty of phytoliths is that they form as casts of the host cells so that many of them have shapes that are peculiar to the plant that made them. Being silica, they remain after dead plant parts rot and can be washed into lake sediments in the runoff water, along with the clays and other minerals.

Phytoliths in tropical soils and sediments afford scientists a means of identifying plants of ancient vegetation that is independent of the

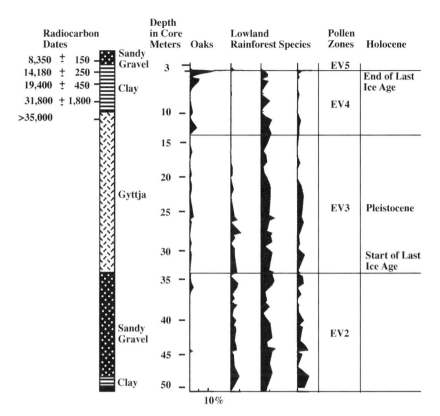

5–2. A pollen diagram of perhaps the past 100,000 years from sediments cored in El Valle (locality indicated on figure 3–4). Similar to the Lake La Yeguada diagram except that time is shown on the vertical axis as depth in meters in the core. Radiocarbon dates are possible only to about 35,000 years, which is reached at 10 meters down the core. This diagram shows that the lowland tropical rain forest species are mostly present throughout the whole glacial cycle. From M. Bush and P. Colinvaux, "A Long Record of Climatic and Vegetation Change in Lowland Panama," *Vegetation Science* 1 (1990): 105–19.

pollen record. Although not all plants make recognizable phytoliths, the ones that do may give a more certain indication that the plant grew locally than do one or two pollen grains that might have blown in.

Like pollen, the phytolith record, particularly at La Yeguada, shows that montane plants grew at low elevation 14,000 years ago. *Magnolia* and sedges of the genus *Carex* were there among the oaks, as was a distinctive phytolith known in modern soils only from highlands.

Thus, the descent of many montane plants by about 800 meters in the Ice Age of Panama seems established and implies that the air at these lower altitudes was colder than it is now. A temperature difference of 800 meters of altitude represents 5–6 degrees centigrade, suggesting that Panama was nearly 6 degrees centigrade colder in the Ice Age than now.

Cold in an ice age seems a reasonable conclusion, and yet it has not been widely accepted because it appeared to violate the results of paleoceanography. For twenty years, oceanographers have proposed that tropical oceans remained warm even as the rest of the world cooled. If the narrow Central American isthmus was bathed in warm oceans it should certainly not have been cooled by as much as 6 degrees centigrade. Maps of past climate used by climate modelers are based largely on reconstructions of past sea surface temperatures. Calculations from the distribution in the sea of foraminifera (microscopic skeletonized, single-celled animals that form part of the plankton) had for several years appeared to suggest that the ocean surface temperatures did not change during the Ice Age, a conclusion difficult to reconcile with isthmian air temperature lowering of 6 degrees. The most recent studies of sea surface temperatures, however, which use the chemistry of corals, have estimated that the surface of the Caribbean Sea in glacial times cooled by 5 degrees centigrade; as large a drop as that suggested for the land by pollen data.

Additional support for the cooling of the tropics comes from well-established studies showing that mountain glaciers in the tropics in Asia, Africa, and the high Andes of Ecuador and Peru descended by 1000 meters or more in the last Ice Age. Ecuadorian rain forests show that cooling occurred in the Amazon basin as well as in the high Andes. The records duplicate the Panamanian histories, showing by pollen, phytoliths, and even chunks of wood that plants now confined to the high mountains descended to rain forest elevations in the last glaciation. Last, oxygen isotopes of pollen in solution lakes in the great limestone terrain of the Petén have recently indicated a glacial lowering of temperature for this region of 8 degrees.

In summary, the whole Central American isthmus, from the Petén of Guatemala to the El Valle caldera in Panama, was cooled in glacial time. Temperatures change constantly over the years, of course, yet it is reasonable to think of the isthmus as having been usually about 6 degrees centigrade colder than now. This appears to have been the dominant influence on climate and vegetation before the start of our warm Holocene interval 10,000 years ago, although locally, as in the Petén and Belize, low rainfall played an important role also. Otherwise, extensive forests appear to have covered the Central American isthmus during the Ice Age and perhaps account for the absence in South America of bison (buffaloes), a late herbivore arrival from the north (see chapter 4).

The Ice Age Vegetation of Central America

The 55-meter core from El Valle appears to span the last inter-glacial and the last glacial cycle until 8000 B.P., a period of about 120,000 years (see fig. 5–2). It thus provides a possible guide to the range of vegetation changes in Central America for the whole of the past 2 million years in which glacial cycles have come and gone. The El Valle core displays five separate pollen episodes or zones, which are based upon a statistical analysis of the composition of all the pollen types as they change through time up the core. Radiocarbon dating of the core back to 30,000 years shows that the two youngest pollen episodes can be correlated to two oxygen isotope stages in the oceans. Isotope stage 1 represents the postglacial warming of the Holocene, or past 10,000 years, and equates to pollen episode EV5, which is typified by a flora like that present before the modern clearings. Isotope stage 2 encompasses the last glacial maximum from 10,000 back to 24,000 and corresponds to the coldest of the pollen episodes, EV4. Older than this there are no independent radiocarbon dates to relate the isotopes to the pollen, but pollen episode EV2 represents a warm period that is most likely equivalent to the warm isotope stage 5 of the last interglacial, which would confirm the age of the El Valle core as about 120,000 years (see fig. 5–2).

The El Valle pollen history, like that at Lake La Yeguada, records the descent of plant species during the glacial periods to a level about 800 meters lower than that at which they grow today. Nevertheless, mixed in with these species at lower levels are species which signal that significant elements of a typically tropical forest never disappear. This is crudely shown by the continued presence of Moraceae and Ur-ticaceae pollen at about 20 percent of the total flora, a datum that stud-ies in the Amazon as well as in Panama have shown means the exis-tence of tropical forest.

To improve on this crude voucher for the presence of tropical for-est, investigations were begun to try to identify the more rare but dis-tinctive pollen grains that give decisive evidence of the presence of tropical rain forest species. Making a reference collection and learning to recognize more than 25,000 Central American tropical species of plants from their pollen are not currently practicable. So pollen traps were put out for a year at a time in vegetation of known composition. The traps caught pollen grains by the thousands, and these were pre-pared for analysis in the same way as pollen from lake mud. They were related fairly easily to one of the plant types of known vegetation from which they had come and for which there existed a complete species list. From this relatively short list, many of the rare pollen types that were not known before were identified, and they turned out to be types that were characteristic of lowland tropical forests. In the Ice Age sediments of El Valle, species such as *Bombacopsis quinata* and *Tricanthera gigantea* and genera like *Luehea, Mortoniodendron, Bocconia, Erythrina, Dipteryx, Bursera, Warszewiczia, Adenocalymna* occurred, all of which appear to be restricted to lowlands below 700 meters in modern Panama.

These data show that tropical forest was maintained within pollen dispersal distance of the El Valle caldera at all stages in a glacial cycle, even though cooling allowed invasion of the crater by trees from the highlands like oaks. The crater floor was the setting of a long battle between the plants from above and those from below. Tropical forest species held ground continuously, perhaps reinforced by plants from the lowlands between the crater and the sea. But species from on high like the oaks were able to take advantage of the 6 degrees centigrade cooling to force their way into the local forests, growing within the crater itself. The 500-meter-high crater floor thus contained hybrid forests like nothing known in modern Central America. When cooling drives plant species down mountains, those that require the most warmth can be expected at the lowest elevation. Tropical forest, therefore, was present in the lowlands in glacial times.

Because the El Valle–Lake La Yeguada pollen history spans all phases of a complete glacial cycle, it gives a general impression of the range of changes, at least at the southern end of the isthmus, throughout the 2 million years of Pleistocene time during which glaciation came and went. Vegetation in the Panama lowlands was always primarily tropical forest, with a mosaic of drier areas as today. Forest always covered the isthmus from sea to sea, but with the cooling of each glacial cycle, parts of the lowland forest were invaded by oaks and other species now confined to the highlands.

Panama also appears to have been a filter for species crossing the land bridge to enter South America. Oak trees, for instance, have had minimal success at making the journey. A number of oak species from the Central American Highlands—from Guatemala to Chiriquí in western Panama—are known, but only one in the Darién. A single oak reached South America proper, first appearing in pollen diagrams from alpine Colombia about 800,000 years ago. Oaks are part of the old flora of Asia and Europe and are representative of plants that would have had to cross the Central American land bridge from north to south.

Glacial cold times would have given oaks their best chance of spreading, yet apparently only one made it across. This would explain why glacial cooling let oaks descend to around the 500-meter contour, as at El Valle, but no lower. The passage from west to east across the Panama lowlands had to be completed by a form of island hopping, from cold upland to cold upland. Given the generally low topography of the Darién, it probably required birds carrying acorns across the intervening patches of tropical rain forest. Even so, in nearly 2 million years of the Ice Age it happened only once.

The Present State of Knowledge

The pollen records from the whole of Central America are still too few to permit drawing of detailed vegetation maps of the past. The strong asymmetry of climate patterns from the Pacific to the Atlantic sides of Central America has meant, for instance, that the Caribbean coast has been subject to the twin effects of cooling and drought with each glacial cycle. Thus, in glacial periods, the rain forests disappeared in such places as Belize and Petén and were replaced by savannas, so the modern rain forests in these areas must be thought of as ecologically young, being no more than 10,000 years old.

There are as yet no long records for the wet Caribbean coasts, from the Mosquitia Coast to the Caribbean slope of the Darién, or for the Pacific coast of El Salvador and Guatemala. The best that researchers can do for these places, for which they have no history, is to apply the generalities of climatic change as they are now known: general cooling on the order of 6 degrees centigrade and reduction in monsoonal rains of 10–20 percent. Regions with marginal climates controlled by monsoons should have been drier: the Petén is a model. Regions with more moisture continued in forest, though with invasions of montane species: El Valle is a model.

From this discussion, it might seem that there is a paradox between the evidence developed for the intercontinental exchange of large groups of herbivores and grazers (described in chapter 4), which apparently requires some development of dry savannas for their passage, and the strong pollen and mineralogical evidence for tropical rain forest and open-water lakes continuing in existence throughout the coldest glacial cycles of the Ice Age in Panama. This is especially so in light of the fact that Bison (buffaloes), the one major group of large herbivores that arrived after the last glaciation, when modern moist forests were clearly reestablished, did not migrate further than Nicaragua en route to South America.

Perhaps the explanation that best reconciles the two sets of data is one that envisages not an open-savanna causeway running through the isthmus but rather a mosaic of plant communities. Tropical forest occupied much of the lowlands but was interpenetrated by novel plant communities from the extensive mid and upper slopes, which included less dense woodland and possibly scrub; the vegetation was a patchwork through which large herbivores could make their way. These glacial plant community mosaics would have changed constantly through succeeding glacial cycles, permitting the migration of different types of animals at different times.

CHAPTER SIX

The Native Peoples of Central America during Precolumbian and Colonial Times

RICHARD COOKE

One of the first things to impress travelers to Central America is the wealth and variety of its Native American heritage: impressive Maya ruins nestled amid the rain forests of the Petén; rows of strange symbols intricately carved in stone; statues of stark deities and head-hunting warriors; exquisite gold, shell, and jadeite jewelry; and today, smoke-filled Indian churches that mingle native and Christian rites; multicol-ored, polyglot markets; and bustling villages of cane and palm thatch.

Behind these visually impressive aspects of native culture lies a long and complex history. Although archaeologists and historians by the hundreds have studied the Maya for more than a century, knowl-edge about the origins of this fascinating culture remains remarkably obscure. South of the Maya lands, where archaeologists are numbered in scores rather than hundreds, large blocks of time and large tracts of space have never been studied.

I shall trace here the history of the native peoples of Central Amer-ica from their arrival some time before 9000 B.C. to the end of the colonial period (1810–21). Attempting to be as geographically and tem-porally impartial as possible, I view the isthmus as a cultural and his-torical unit characterized by a subtle and protracted interaction among indigenous and nonisthmian peoples, plants, and ideas.

Paleoindians

The oldest artifacts from Central America are stone tools made by Paleoindians, hunters and gatherers who lived at the very end of the last Ice Age. Some archaeologists believe that these Paleoindians were the first Americans and thus ancestors of all native peoples who survived the European conquest south of Mexico. Entering the New World about 12,000 years ago, they would have crossed a land bridge between Siberia and Alaska that was exposed by lowered sea level. They then migrated rapidly southward, causing or hastening the extinction of many large mammals that had been perhaps already reduced in number by drastic environmental changes during the Ice Age. Most archaeologists think that the Paleoindians were preceded by earlier immigrants, identified at a few South American sites but not yet discovered in Central America.

No Central American Paleoindian sites have been dated by the radiocarbon method. Artifacts and animal bones have not yet been found together. Paleoindian stone spear points, skin scrapers, and drills, however, are very similar to examples dated in North and South America between about 9500 and 8500 B.C. One specialized type of spear point was whittled down bilaterally from a piece of glassy stone and had one or two flutes removed vertically from the base. The flutes facilitated hafting of the point to a bone or wood foreshaft. Fluted points found in Belize (Ladyville), Guatemala (Los Tapiales), Costa Rica (Turrialba, Lake Arenal), and Panama (La Mula, Lake Madden) are similar to those made by North American Clovis mammoth-hunters (fig. 6–1A). Another variety, called fishtail (fig. 6–1B) because of its narrow eared stem, has been found in Costa Rica (Turrialba), and Panama (Lake Madden) (figs. 6–2, 6–3). Fishtail points occur in South American sites alongside ground sloth, mastodon, and horse remains. These extinct terrestrial mammals, as well as others mentioned in chapter 4, were surely hunted by Central American Paleoindians.

Although the coldest and driest part of the last Ice Age was over by Paleoindian times, temperatures and rainfall were still lower than they are today. Central America was becoming more forested, but with vegetation types that do not have modern parallels. Oak woods were replacing temperate scrublands in lowland Guatemala and Belize. In Costa Rican and Panamanian foothills, oak-magnolia woods were beginning to move up slope as lowland tropical species moved in. Along the Pacific plains, the vegetation mosaic was probably quite open in places. Paleoindians roamed widely through these habitats, from above the tree line down to sea level. Tool kits differ little from site to site, suggesting that widely scattered bands were constantly on the move.

A climate like that of today was first established in Central America about 8500 B.C. The descendants of the Paleoindians crafted new

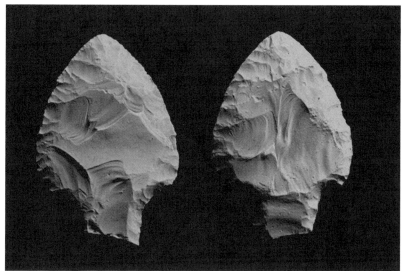

A

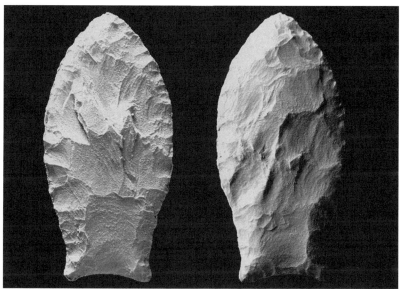

B

6–1. Paleoindian fluted spear points from Lake Madden. Found at the surface, they probably date from about 9500 to 8500 B.C.: (A) fishtail variety, whose stem was broken in antiquity; made of brown jasper, 59 millimeters long; (B) Clovislike variety; yellow jasper, 76 millimeters long. Photographs by Junius B. Bird.

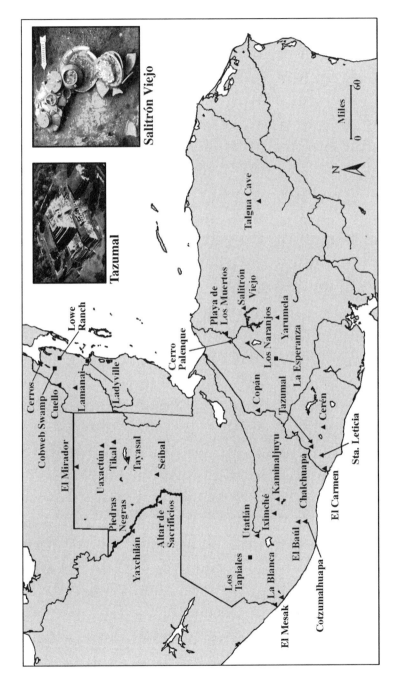

Tazumal

Salitrón Viejo

6–2. Principal archaeological sites in northern Central America, that is, Chiapas state in Mexico, Belize, Guatemala, El Salvador, and Honduras. Squares locate Paleoindian sites; triangles villages and towns with ceramics. (*Left inset*) : The Maya center of Tazumal in El Salvador (A.D.600–1200). Photograph by Payson Sheets; (*right inset*) : Salitrón Viejo, Honduras, a Late Preclassic Maya burial site with Usulután decorated bowls. Photograph by George Hasemann.

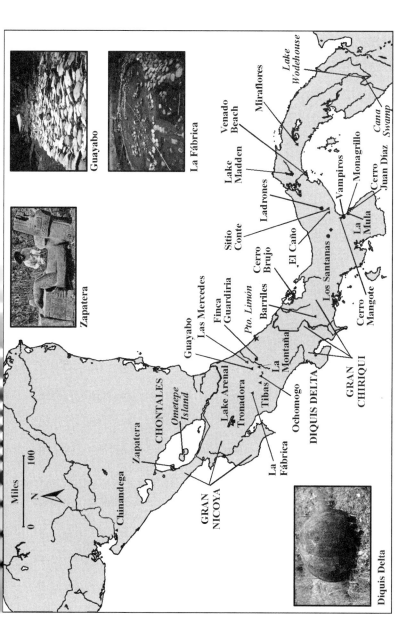

6–3. Principal archaeological sites in southern Central America, that is, Nicaragua, Costa Rica, and Panama. Squares locate Paleoindian sites, circles Preceramic sites, and triangles towns and villages with ceramics. (*Bottom left inset*): Stone ball, 1.7 meters in diameter, from Sitio Cansot, Diquis Delta, Costa Rica, A.D. 600–1000. Photograph by Ifigenia Quintanilla. (*Top right insets*): Zapatera, Nicaragua, Mexican-style stone sculpture. Photograph from S. K. Lothrop, *Pottery of Costa Rica and Nicaragua*, vol. 1, plate V (Museum of the American Indian, Heye Foundation, 1926); Guayabo, Costa Rica, stone causeway, A.D. 900–1500. Photograph by Federico Solano; La Fábrica, Costa Rica, round house structure, A.D. 500–800. Photograph by Juan Vicente Guerrero.

kinds of unfluted spear points with serrated edges and lateral barbs. They would have used these for hunting white-tailed deer, peccaries, and other swift-footed, secretive mammals that had avoided extinction. Their distribution in central Panama and the Quiché Basin of Guatemala suggests that human populations were growing in these regions, which have been well surveyed by archaeologists. They were staying for longer periods at single sites and were dedicating more time to collecting and preparing plant foods. One of these was arrowroot, which was planted in the foothills and along the Pacific coast of Panama by 6500 B.C., near sites like Los Vampiros (fig. 6–4A). Its small, starchy tubers may have been mashed with small cobble grinding stones with flattened edges (fig. 6–4B).

Plant Domestication and Agriculture

Arrowroot may have crossed the subtle frontier between a wild and a cultivated state within the confines of the isthmus, but most of the plants that by the time of Spanish contact (A.D. 1502) had become staple foods and utility plants—maize, squash, beans (lima, jack, and kidney), manioc or cassava, sweet potatoes, a native yam (*Dioscorea trifida*), agave or sisal, and cotton—were first cultivated elsewhere by Native Americans. Maize's domestication history is by far the best known. Its ancestor is Balsas teosinte, a wild grass native to the midelevation valleys of southwestern Mexico. It had moved or been taken into Central America by 5000–2500 B.C.

Logically, human population size and density and relative social complexity were closely related to the nature of production, storage, and dispersal of plant foods and materials. The magnificent Maya culture is inconceivable without the huge food surpluses required to sustain porters, masons, sculptors, stucco workers, stone tool experts, wood-carvers, painters, laborers, priests, curers, and diviners, as well as to barter for the vital but locally scarce commodities that such a heterogeneous society demanded. To achieve these surpluses, the Maya developed *intensive* methods of agriculture: ditching, draining, mounding, terracing, and fertilizing. There is ample evidence of such practices in Belize, El Salvador, and lowland Guatemala (fig. 6–5).

Not all parts of Central America, however, attained this degree of agricultural specialization. In particularly humid zones, where cut vegetation cannot be burned every year, plants were cultivated in small multispecies plots that were fallowed for many years before being reused. Farmers constantly moved their villages to be near new fields.

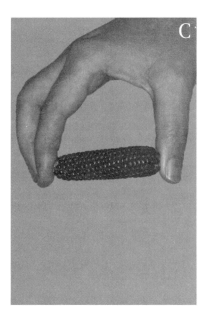

6–4. Early plant cultivation: (A) arrowroot plants (*Maranta arundinacea*) , which were cultivated about 6000 B.C. Photograph by Richard Cooke; (B) cobble mashing stones for root crops, Casita de Piedra, Chiriquí, Panama, 4600–2300 B.C. Photograph by Anthony J. Ranere; (C) modern Argentine popcorn. Photograph by Dolores Piperno; (D) Cueva de los Ladrones, a rock-shelter in Panama. Photograph by Junius B. Bird.

6–5. Evidence for Maya agriculture preserved under ash from a volcanic eruption. A portion of a maize field, furrowed and with holes left by maize plants, is revealed beneath debris from the eruption of Laguna Caldera, about A.D. 600 at Joya del Cerén in El Salvador. *(Left inset):* a ceramic vessel with squash seeds; *(right inset)* : sisal cordage. Photographs by Payson Sheets.

An interesting Precolumbian example of this itinerant forest-farming lifestyle, whose origins date to 6500 B.C., is Cerro Brujo (A.D. 600–900) on the wet Caribbean lowlands of Bocas del Toro, Panama (see fig. 6–3).

Archaeologists have had difficulty finding Central American sites that belong to the initial stages of cultivation, that is, before large villages existed and before pottery was in use. To determine when Central America's most important crops appeared, how each one developed in size, shape, and adaptability, and when and how they were cultivated, archaeologists seek collaboration with other specialists. Paleoecologists reconstruct regional vegetation histories and identify the first appearance of specific cultigens by studying pollen, phytoliths, and charcoal deposited in lake sediments (see chapter 5). Specialists in isotope geochemistry infer changes in diet by calculating the proportions of nitrogen and carbon isotopes present in ancient human bone (the heavy consumption of maize as opposed to that of root crops leaves very different isotopic signals). Botanists and geneticists reconstruct the evolution of domesticated plant species by studying the molecular biology and morphology of present-day varieties and comparing these with ancient carbonized fragments painstakingly recovered from archaeological sites.

Data from the Pacific slopes of central Panama illustrate the importance of multidisciplinary research in reconstructing the slow evolution of plant domestication in Central American forests before 2000 B.C. and the rather rapid expansion of slash-and-burn agriculture after this date. Lake La Yeguada (see fig. 3–4), located at 650 meters above sea level, has provided a 14,000-year record of vegetation history. Paleoindians first camped and lit fires here about 9000 B.C., when the neighboring foothills were still covered with montane forest. As conditions became wetter and warmer, people came more frequently to the lakeside. Their annual cutting and burning encouraged plants that like gaps in the forest: palms that produce bunches of oily, protein-rich fruits and herbs and vines such as arrowroot that store nutrients in tuberous underground organs. Between 5000 and 3000 B.C. the vegetation shows signs of having been continually cut and burned for the planting of crops. Maize pollen and phytoliths appear in the lake sediments between 3000 and 2000 B.C. and at nearby rock-shelters, for example, Los Santanas and Los Ladrones (see fig. 6–3), between 5000 and 2500 B.C. Farming activities around Lake La Yeguada peaked about 2000 B.C., but declined soon after, probably because the surrounding countryside could no longer sustain a growing population with long-fallow, slash-and-burn methods (the effects of farming on the La Yeguada vegetation are summarized in figure 5–1).

Other regions of Central America have not yet provided evidence for the *continual* post–Ice Age Native American modification of tropical forests evident near Lake La Yeguada. Lake core records, however, point to slash-and-burn activities and to the use of maize by 3000–2000 B.C. In some areas, farmers colonized previously unpopulated or thinly populated forests quickly. This happened in the Petén region of lowland Guatemala, around Lake Yojoa near the famous Maya city of Copán in Honduras (fig. 6–6B), as well as in northern Belize, where Cobweb Swamp yields evidence of low-scale or sporadic farming activities about 3000 B.C., by which time maize and manioc were being planted. Some 500 years later, large areas of forest had been burned and cleared, probably by ancestors of the Maya.

In the Caribbean lowlands of central Panama (Lake Gatún; see fig. 3–4), low-scale farming activities were likewise followed by the rapid expansion of slash-and-burn and maize cultivation during the third and second millennia B.C. In the Tuira Valley of Darién Province in Panama, where no archaeological fieldwork has been undertaken this century, sediment records from Lake Wodehouse and the Cana swamp suggest continual but irregular slash-and-burn farming, with maize, between

6–6. Two famous Classic Maya towns: (A) Tikal, Guatemala, Temple I ("Giant Jaguar"), which stands above Ruler A's tomb and dates from about A.D. 700; (B) Copán, Honduras, view of Structure 10 (ball court). Photographs by permission of American Museum of Natural History.

2000 B.C. and A.D. 1600, when the Spanish destroyed the native population, and forests returned to the area.

The varieties of plants cultivated in Central American forest plots before the advent of settled village life were much smaller and less productive than the varieties grown by contact times. Maize, for example, was probably a small-eared, low-rowed popcorn, phytoliths of which have been found in human refuse deposits at Los Ladrones and other Panamanian rock-shelters dated between 5000 and 1000 B.C. (see fig. 6–4C, D). The meager list of plant species that have been recorded for these early (pre-2000 B.C.) Central American farmers—maize, arrowroot, manioc, squash—is surely a poor reflection of the total numbers of plant species that were manipulated by humans during this long period of experimentation. To judge from the large numbers of carbonized palm and other fruits in pre-2000 B.C. archaeological sites, wild plant resources still played an important part in the regional diet. House gardens, in fact, remained an integral part of Central American subsistence until the conquest: not neat, tidy ones with straight rows and geometric beds, but unruly mixtures of herbs, shrubs, vines, and trees that were used for food, dyes, condiments, stimulants, colorants, incense, and medicines. Important trees and shrubs were sometimes planted in groves: coyol palms (*Acrocomia*) for wine and thatch; sapotes, *mameyes,* nances, and breadnuts for fruit; cotton and sisal for clothes and hammocks; copal (*Hymenea courbaril*) for incense; and cacao. By Aztec times, cacao had become a valuable commodity. The beans were used as currency for bartering. Chocolate, sometimes flavored with hot peppers, was a high-class drink.

Harvesting the Sea

Considering how rapidly they destroyed native agriculture, it is ironic to read how enchanted Spanish soldiers were by the verdure of Central American towns and villages. They were also impressed by the productivity of the two contrasting oceans that bathe Central America's coasts and by the opportunities they afforded for warfare and trade. Christopher Columbus's morale picked up when he spied large canoes transporting well-dressed merchants up and down the Caribbean coast in search of gold trinkets and cacao beans. His son, Ferdinand, marveled at the ingenious methods used by Veraguan fisherfolk in Panama to catch shoals of tiny fish that annually entered coastal rivers. Fifteen years later, on the opposite coast of Panama, the infamous Lieutenant

Espinosa was overawed by what he described as tons of fish hauled with sisal nets from Chame Bay.

The earliest reliable evidence for coastal settlements in Central America comes from Parita Bay, a mangrove-fringed embayment on the Pacific coast of central Panama. (Huge shell mounds in the Caribbean lagoon district of Nicaragua may date beyond 5000 B.C., but they have not been excavated by professional archaeologists.) Some early coastal sites may lie buried under recent alluvium (river-laid sentiment). Others may have been missed by incomplete archaeological surveys. Even so, archaeologists are puzzled as to why ecologically and topographically similar estuaries further up the Central American Pacific coast were apparently not exploited regularly by Native Americans until several millennia later than Parita Bay.

The best-known Parita Bay site is Cerro Mangote (5000–3000 B.C.), a camp or house cluster situated on top of a long hill. It is now 8 kilometers inland (see fig. 6–3) but at that time was considerably closer to the sea. The delta of the nearby Santa María River had not yet formed. Archaeologists have found ninety human skeletons placed within kitchen refuse. Bodies were first left to decompose, and then the bones were collected, painted with red ochre, wrapped in cloth or basketry, and carefully redeposited as packages containing several individuals (fig. 6–7A). Such secondary burial appears to have a long history in Central America. It has been reported from the Talgua Cave in Honduras, where probably non-Maya people buried their dead with pottery, jade, and marble offerings about 1000 B.C. (fig. 6–8B). The custom was still practiced at Cerro Juan Díaz in central Panama about A.D. 500 (fig. 6–7B). It survived among the Bribri, Cabécar, and Ngöbé peoples of Costa Rica and Panama until recent times.

Cerro Mangote's occupants stayed long enough to plant gardens. Stable isotope geochemistry indicates that they consumed some maize. They also collected shellfish and crabs, trapped birds (especially white ibis), and hunted iguanas, raccoon, white-tailed deer, anteaters, rabbits, and pacas. On occasion, they were lucky enough to seize sea turtles. But, most of all, they were attracted to the fishing of abundant sea catfish, snook, grunts, croakers, toadfish, and small jacks. The bones of these fish appear not only in Cerro Mangote's middens, but also at small farming settlements occupied by culturally related people more than 20 kilometers from the coastline. This suggests that dried and salted fish were being transported inland—another long-lived custom in Parita Bay. Modern studies of fish distribution and fishing technology show that nearly all the fish species consumed at Cerro Mangote can be

6–7. Mortuary customs: (A) Cerro Mangote, central Panama, secondary burials in packages dating from about 4000 B.C. Photograph by Anthony J. Ranere; (B) Cerro Juan Díaz, central Panama, secondary burials in packages dating from about 500 B.C. Photograph by Luis Alberto Sánchez.

A

B

6–8. Mortuary customs: (A) Ochomogo, Cartago, Costa Rica, stone graves from about A.D. 800–1500. Photograph by Aida Blanco; (B) Talgua Cave, Honduras, 1000–750 B.C., multiple burial bundles now encrusted with calcite crystals. Photograph by George Hasemann.

A

B

C

6–9. Studies of modern fishing in Parita Bay help archaeologists to interpret the remains of fish bones discovered in nearby sites: (A) stationary pole and net trap, 1993. Photograph by Laurel Breece; (B) salted leather jackets (*Oligoplites altus)* hanging up to dry at Monagrillo, Panama. Photograph by Richard Cooke; (C) *Cathorops tuyra,* a freshwater-tolerant catfish species common in the Precolumbian middens of Parita Bay. Photograph by Gerald Allen.

caught with pole-and-net traps placed on the estuarine mudflats (fig. 6–9A, B, C). This so-called actualistic research does not *prove* that Cerro Mangote's inhabitants fished in this way. But the use of tidal traps would explain why no hooks, harpoons, or net weights have been found at early Parita Bay sites, in spite of the dietary prominence of marine resources.

Cerro Mangote is called a Preceramic site because it lacks pottery. Many Central American archaeologists feel that a farming way of life is inconceivable without pottery and, indeed, that people without pottery are people without culture! But societies growing maize and practicing slash-and-burn cultivation precede the appearance of pottery in Central America by two thousand years. Presumably, these Preceramic hunter-fisher-cultivators cooked and stored food and liquids in vessels made out of perishable plant materials, like tree- and bottle-gourds. Their pots and pans may have been wild banana leaves (*Heliconia*), which native peoples still use for roasting foods over a fire.

The Appearance of Pottery

The oldest pottery in Central America is called Monagrillo after the Parita Bay shell mound where it was first discovered (see fig. 6–3). Known only from central Panama, it appears about 3000 B.C. and lasts until about 1000 B.C.. At some rock-shelters, it rests upon Preceramic strata. Ceramics were manufactured several millennia before this in Brazil and Colombia, and it is possible that the *idea* of making clay vessels came from there. Stylistically, however, Monagrillo pottery has no counterparts. Its introduction is not accompanied by sudden changes in other aspects of life. It appears to be a local ware manufactured with local clays. Its slapdash bowls and collarless jars were made for two thousand years and are just what one would expect itinerant fisher-farmers to fashion (fig. 6–10A, B, C).

Farther north, from Costa Rica into Guatemala, pottery does not appear until about 2000–1500 B.C., but it is generally much better made than Monagrillo ceramics, with raw materials that were not always available in the vicinity. Decoration consists of incised, impressed, and stamped patterns often arranged in geometric zones separated by bands of highly polished red paint (fig. 6–10F).

Soconusco

One of the earliest centers of the more refined northern ceramic tradition was the narrow, fertile strip of lowlands that straddle the Mexican state of Chiapas and Guatemala, a region called Soconusco in colonial times. Its deep, rich soils were well suited for maize, beans, and squash, which by this time were similar to modern varieties, although smaller. Extensive mangroves and estuaries made excellent fishing and shellfishing grounds.

The oldest Soconusco villages, dated about 1700–800 B.C., consisted of whitewashed, pole-and-daub houses built on low platforms and strung out along river courses near the coast. Gradually, some set-

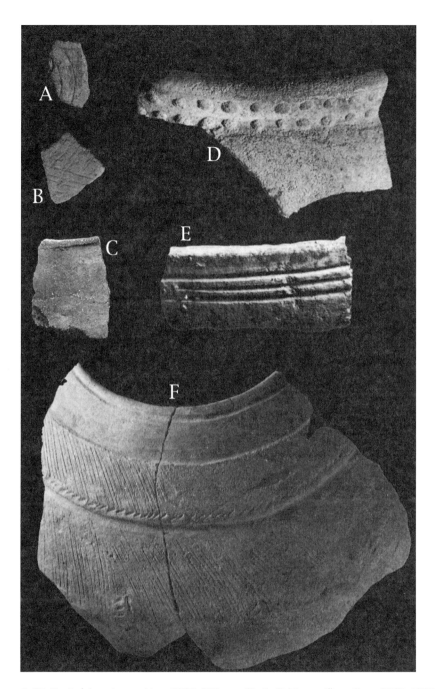

6–10. Central American pottery, 3000–500 B.C.: (A, B, C) Monagrillo pottery, 3000–1200 B.C., central Panama. Photograph by Richard Cooke; (D, E) La Montaña pottery, Costa Rica, 1000–500 B.C.. Photograph by Maritza Gutiérrez; (F) Cuadros pottery, La Blanca, Pacific coastal Guatemala, 1100–500 B.C.. From M. D. Coe and K. V. Flannery, "Early Cultures of Human Ecology of South Coastal Guatemala," *Smithsonian Contributions to Anthropology* 3 (1967): plate 9g.

tlements became larger and more important than the rest (fig. 6–11). They were endowed with raised ceremonial areas, a few large buildings—perhaps the residences of chiefs or congress halls—and mounds 25 meters high. Typical examples are La Blanca, El Infierno, La Zarca, and El Mesak in southwest Guatemala. Jewelry made of jade, greenstone, and schist bears witness to a growing upper-class demand for overt symbols of rank.

The Olmec Influence

Formerly, these social developments were thought to be due, directly or indirectly, to the influence of the Olmec, whose culture reached its zenith in the humid lowlands of the Mexican states of Veracruz and Tabasco. Between 1200 and 800 B.C., the Olmec built earthen mounds, jaguar-faced pavements, and colossal human heads carved out of basalt. A once popular theory postulated that they migrated into western Mexico and Guatemala and introduced maize agriculture. In the process, they displaced or absorbed earlier, simpler cultures who used wooden grater-boards studded with tiny obsidian teeth to grate a "bitter" variety of manioc that needs to be washed and squeezed in order to dispel harmful toxins.

It is now known that maize appeared in Central America long before 1200 B.C. Bitter manioc was not used in Central America in contact times but was spread by Caribbean islanders and Spanish seafarers after the conquest. Olmec migration and domination theories have now given way to more elaborate models to explain the distribution of Olmec-like objects. These stress social interaction and economic integration over a large swath of territory that stretched from the central Mexican highlands (Tlatilco, Chalcatzingo) and the Gulf Coast (San Lorenzo, La Venta), through Oaxaca (San José Mogote), Soconusco, the Chiapas and Guatemalan Highlands (Kaminaljuyú), and into El Salvador (Chalchuapa, El Carmen). The peoples of this territory shared a belief system whose most striking icon is a cleft-headed jaguar-human creature with a noticeably drooping mouth. It appears on a wide variety of small and large objects: ceramic vessels, jade and serpentine statuettes and axes, cave wall paintings, and bas-reliefs (pl. 9A, C). Rotund, pot-bellied sculptures belong to the same iconographic tradition (pl. 9B). During the first millennium B.C., some towns in this economic sphere became large and influential, for example, Kaminaljuyú, in the Guatemalan highlands; Izapa, just across the Mexican border; El Baúl and Cotzumalhuapa, on the Pacific coast of Guatemala; and Chalchuapa and El Car-

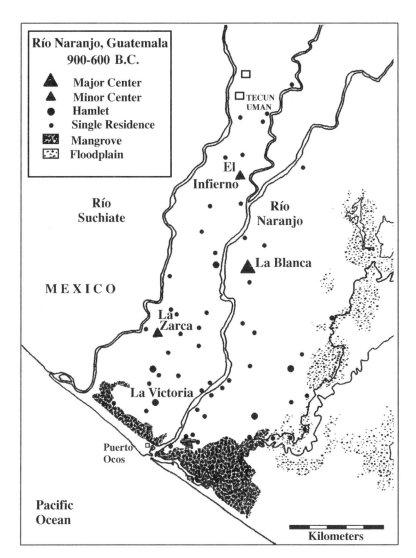

6–11. Evidence of increasing social complexity: a survey of the pattern and hierarchy of settlements and residences in the Río Naranjo valley, southwestern Guatemala, demonstrates notable differences in site size and status between 900–600 B.C. Modified from Michael W. Love, "Style and Social Complexity in Formative Mesoamerica," in *The Formation of Complex Society,* William R. Fowler, Jr., ed. (CRC Press, Boca Raton, Fla., 1991), fig. 6.

men in western El Salvador. Kaminaljuyú was endowed with impressive public monuments alongside which stone stelae were erected. The symbolism and glyphic writing of the stelae provide an interesting link between Olmec and later Maya cultures (fig. 6–12). Kaminaljuyú's wealth was probably related to the production and distribution of obsidian (volcanic glass) tools. Another important center was Chalchuapa, where obsidian and Usulután pottery were important trading commodities. Usulután vessels, which were decorated with two coats of suspended clay to improve their impermeability, were traded far and wide in Central America (pl. 10B). Nature's hand dealt a fatal blow to Chalchuapa's regional dominance: a massive eruption of Ilopango around A.D. 250 led to its abandonment and, some archaeologists believe, to the exodus of the survivors to the eastern Maya lowlands.

The Origins of the Maya

Expanding populations and larger, planned settlements are also visible in Belize toward the end of the second millennium B.C., when peoples that can now be called early Maya shifted emphasis from slash-and-burn hillslope farming to wetland and riverine agriculture specialized on root crops and maize. At Cuello and neighboring sites they built houses with lime-plaster floors. They kept dogs for food, fished for freshwater turtles, and hunted secondary growth and forest edge animals, especially the white-tailed deer. They imported blue jade for making beads from the Motagua Valley (Guatemala) and Guerrero state (Mexico).

There are few geographical barriers in the eastern lowlands of northern Belize and Guatemala. Quarreling groups in early Maya villages could readily split off and move to uninhabited territory. This fissioning was responsible for the spread of a distinctive monochrome and red-daubed pottery called Mamom, which includes standardized human figures with glossy or waxy surfaces. By the beginning of the Christian era, however, the eastern lowlands had essentially filled up with people. There was a huge demand for such raw materials as salt, obsidian, chert, and abrasive stones for making maize-grinding implements, which in this mostly limestone country are found only at a few, widely scattered localities. The early ascendancy of such well-known Maya towns (see fig. 6–2) as Tikal, Seibal, El Mirador, Uaxactún (Guatemala), Lamanai, and Cerros (Belize) was undoubtedly linked to the control of the production and distribution of these materials and of artifacts made of them.

6–12. Early stelae art at Kaminaljuyú, Guatemala. This stone depicts a human figure wearing the mask of the "long-nosed god" and holding a serpent supported by the mask of a "snub-nosed dragon." The height of the piece is 109 centimeters, and it dates from about 400 B.C. From Doris Z. Stone: *Precolumbian Man Finds Central America* (Peabody Museum of Archaeology and Ethnology, Harvard University, Cambridge, 1972), figures on p. 72.

In spite of abundant rainfall, the Petén and Belize have intense dry seasons. Many areas are characterized by a stark limestone landscape, where drainage is subterranean. Lakes, swamps, and wells are vital for agricultural productivity and stability. It is likely that access to and control of the water supply were related to the emergence of an upper class, who invested their wealth in elite residential zones and sanctified their power by building elaborate temples. At some Petén sites—for example, Uaxactún and Tikal—archaeologists have found, underneath later, larger structures, small stepped pyramids, built during the first millennium B.C. Their stucco-faced stairways are flanked by monstrous jaguar-masks.

"Northern" and "Southern" Peoples

By the end of the first millennium B.C., native society in Central America was taking very different paths toward cultural complexity. The northern sector was from now on characterized by literate, history-conscious peoples who lived in planned settlements, built temples and pyramids around courtyards, played games with rubber balls, erected stone stelae with human figures engraved or sculpted upon them, and recorded on stone, wood, and deerskin details about kingship, warfare, and astronomy. Building upon knowledge transmitted by earlier, Olmec-inspired cultures, they developed a remarkable method of recording the passage of time that was based on the interdigitation of a *solar* and a *ritual* calendar. The solar year comprised 20 months of 18 days each (360 days), to which was added a 5-day year-end period called the *Uayeb*. The Uayeb was considered to be very unlucky. The ritual calendar consisted of 13 named days arranged sequentially in 20 groups. Thus, 260 days elapsed before a particular day came round again. The two calendars ran concurrently, like cogwheels. The initial named days of each calendar met up every 18,980 days, or 52 years. This portentous event inspired frenzied religious and building activity.

The southern sector was less urbanized than the northern, less literate (or nonliterate), had fewer and less rigid social classes, and promoted what can be regarded as a more personalized concept of art and religion. Some regions, eastern Honduras and highland Costa Rica, for example, developed quite complex urban architecture. But in general, social and economic differences among people and sites were symbolized more often by exquisitely fashioned portable objects of gold, shell, wood, and bone than by monumental statuary and buildings.

The boundary between the more developed northerners and the less developed southerners starts in the archaeological region called Gran Nicoya (see fig. 6–3), runs along the western shore of Lake Nicaragua, and then northward to the Honduran coast west of the Bay Islands. Like many frontiers, however, this one varied in space and time. The great Maya towns stretched south as far as central El Salvador (Tazumal, Cerén) and westernmost Honduras (Copán) (see fig. 6–2 insert and fig. 6–6B). Contact with them was probably the motor that drove such vibrant mercantile towns in Honduras as Los Naranjos, Cerro Palenque, Playa de los Muertos, and Yarumela. Their architectural and cultural development parallels closely that of the preclassic and Classic Maya (pl. 11F). Perhaps they were multicultural societies peopled by both Maya and speakers of Misumalpan and Paya-Chibcha languages.

After A.D. 750, the fall of the great Mexican metropolis of Teoti-huacán was followed by ethnic and political turmoil akin to that which followed the collapse of the western Roman Empire, when Germanic and Slavic tribes engaged in long migrations in search of land and booty. Peoples who spoke languages of the Nahua family, to which Aztec belongs, initiated migrations from Mexico into Central America, in some cases traveling well south of the Maya lands. The Pipil settled in southeastern Guatemala and El Salvador west and south of the Lempa River and subsequently exerted strong influence upon some re-gions of Honduras and Nicaragua. Other speakers of Nahua languages—the Chorotega (or Mangue), the Subtiaba, and, later, the Nicarao—established towns between the Gulf of Fonseca and the tip of the Nicoya Peninsula (see figs. 3–3, 3–4). The Spanish found Nahua-speaking peo-ples at the mouth of the San Juan River, which flows out of Lake Nicaragua into the Caribbean, and further down the coast at the Costa Rica–Panama border. The so-called foreigners they observed near Nom-bre de Dios in Panama may also have been Nahua speakers.

Some archaeologists have claimed that migrations also occurred in the opposite direction—from South into Central America. Although it is intellectually untenable to link a specific set of Precolumbian arti-facts or sites to a particular surviving modern language, it is likely that the Precolumbian inhabitants of much of the southern half of Central America were an ancient population, not immigrants who appeared in the last 1000 to 2000 years. Ethnohistory (the study of contact-period documents), historical linguistics, and genetics suggest that the twelve ethnic groups that still speak languages belonging to the Misumalpan, Paya-Chibchan, and Chocoan families and that live between Caribbean Nicaragua and Panama represent only the latest stage of a protracted so-ciopolitical fragmentation of an ancestral population that has lived in, or very near, their present-day locations for many thousands of years (10,000 years, say the mitochondrial DNA specialists with regard to the Central American speakers of Paya-Chibcha languages!).

The Later Maya

Historical and geographical continuity is also manifest in the Maya world. Millions of people in northern Central America, southern Mex-ico, and the Yucatán still speak one of several languages in the Maya family. Archaeologists try to relate patterns in the distribution of Pre-columbian architecture and artifacts to particular Maya languages. For example, some scholars believe that Maya sites in El Salvador were occupied by speakers of the Chol language. The nearer an archaeological culture is to the conquest, the more objective the correlations become because Spanish chroniclers and priests and a few Spanish-speaking Mayans wrote down Maya history while it was fresh in their memories. Clearly, the Maya structures now in ruins were built not by Egyptians or Phoenicians but by Mayans.

Lowland Maya culture reached its zenith during the so-called Classic period, which began about A.D. 300 and lasted until about A.D. 900. During this period, the Maya wrote down dates in the Long Count, which traces events from a base date, 3113 B.C., akin to the mythical date for the foundation of Rome. To count years, the Classic Maya (fig. 6–13) tallied *baktun* (periods of 144,000 days), *katun* (7200 days), *tun* (360 days), *uinal* (20 days), and *kin* (1 day). The nasty five year-end days belonged to the *Uayeb.* A date written 8.12.14.8.15, for example, meant that 8 *baktun,* 12 *katun,* 14 *tun,* 8 *uinal,* and 15 *kin* had elapsed since 3113 B.C. This totals 1,243,615 days, or, dividing by 365.25, 3405 years, which, when subtracted from 3113 B.C., gives A.D. 292—the oldest Long Count date so far recorded for the eastern Maya lowlands. The Maya wrote numbers as dots [1] and bars [5]. They had a symbol for zero, too (fig. 6–13).

Classic Tikal, which covered more than 120 square kilometers, is reckoned to have supported 65,000–80,000 people during Ruler A's energetic and expansive rule, which ended in A.D. 731 (see fig. 6–6A). This is less than the estimated population of coeval cities in the Mexican highlands, for example, Teotihuacán. The urbanized area was also less built up. Individual houses or clusters of houses were surrounded by gardens, some almost a hectare in size. Geographers identify three ranks of ceremonial and civic centers as well as rural settlements in outlying areas, where poor people lived far removed from the bustle and ceremony of the great pyramid centers.

As noted earlier, the Maya were agricultural peoples who developed intensive methods of cultivation. The annual agricultural round exerted a strong influence on their belief systems and calendar. Maize, for example, is represented as a long-haired young man (see pl. 10A). Power, curing, and astronomical prediction seem to have gone hand in hand. Painted Maya murals and ceramics depict gaudy, noisy ceremonies that involve bloodletting, ritual sacrifice of people and animals, and masked dancers. Caves, often with decorated walls, were important sacred precincts.

Maya hunting and fishing show strong regional preferences and strong class biases. Some communities liked dogs and turkeys more

6–13. Classic Maya writing (glyphs). Numerals one through four are indicated by dots; five and multiples by bars; there is a special symbol for zero. Symbols, or glyphs, are also shown for the various units the Maya used to count the passage of time. Examples of the glyphs of a ruler (Bird Jaguar) and a city (Copán) are also shown. Modified from Muriel Weaver Porter, *The Aztecs, Mayas and their Predecessors* (Academic Press, Orlando, Fla., 1981), figs. 8, 15, 19.

than others. Riverside cities like Altar de Sacrificios and Seibal (see fig. 6–2), in the far western Petén, did not exploit all the aquatic resources available to them, as though some were tabooed or low-class foods. Upper-class people consumed the largest quantities of meat, especially deer. In fact, the gradual contraction of wooded and second-growth habitats may have convinced the Maya to *manage* deer by herding and corralling. Feasting was very important everywhere, and special hunts were arranged to provide enough meat to satisfy guests (fig. 6–14). Dressed deer carcasses were sent out to islands off Belize and the Yucatán.

The Maya political scene was like that of pre-Macedonian Greece or Renaissance Italy. The elites of the most important towns had a profound sense of dynastic history, and many of the famous carved stelae record the births, accessions, deaths, and defeats of named rulers. Rulers and towns had their own "badges," or name-glyphs, so that epigraphers can recognize the conquest of one city or dynasty by another (see fig. 6–13). A lintel at Yaxchilán, on the Mexican bank of the Usumacinta River, records the capture of Jeweled Skull by Bird Jaguar, who was lord of Yaxchilán at the end of the seventh century A.D. (fig. 6–15).

Changes in the dress and personal adornment of prominent figures on stelae and murals inform us about political contacts between the lowland Maya and neighboring cultures. For example, some archaeologists have proposed that Curl Snout, ruler of Tikal, was either from Teotihuacán or Kaminaljuyú in the Guatemalan highlands, where Teotihuacanos may have been in charge of obsidian tool production and ex-

6–14. A Maya deer ceremony. Deer and peccaries (wild piglike animals) were closely associated with agriculture in the Maya mind. This scene is from a polychrome vase. Men brandishing spears remove a stag's antlers; a vulture hovers overhead, and to the right is a sacred tree with a snake curled around the trunk. The scene in general symbolizes water, rain, sacrifice, and agricultural security. From M. Pohl and L. Feldman, "The Traditional Role of Women and Animals in Lowland Maya Economy, 286–299," in K. Flannery, ed., *Maya Subsistence* (Academic Press, Orlando, Fla., 1982).

6–15. Bird Jaguar captures Jewelled Skull. Bird Jaguar became ruler of Yaxchilán in A.D. 752 and is identified by the second glyph down on the top right. Jewelled Skull is identified by the glyph on his thigh. From Norman Hammond, *Ancient Maya Civilization* (Cambridge University Press, New York, 1982).

port. His successor Stormy Sky, whose reign started about A.D. 445, is accompanied on stelae by two shield-bearers dressed like Mexicans (fig. 6–16). One of the shields is decorated with the face of the Mexican rain god, Tlaloc, who also appears on polychrome pottery from Gran Nicoya (see pl. 11E). Perhaps Curl Snout and Stormy Sky were both Maya but liked to show off Teotihuacano ways in the manner of many Indian rulers during the British raj.

The decline and fall of Teotihuacán, at that time the largest city in the world with perhaps 200,000 inhabitants, was intimately related to the demise of many of the eastern lowland Classic Maya cities between roughly A.D. 700 and 900. Militaristic monuments at Piedras Negras and Yaxchilán, on the border between the Mexican state of Chiapas and Guatemala, suggest that buffer states may have been set up, rather like

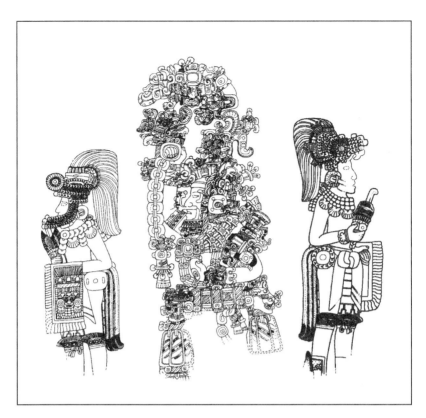

6–16. Mexican influences at Tikal, Guatemala. Curled Nose is flanked by two warriors in Mexican dress and arms. From Norman Hammond, *Ancient Maya Civilization* (Cambridge University Press, New York, 1982).

those that usurpers to the Roman throne established at the edges of the Empire to palliate the chaos of the barbarian incursions. These changes in the regional power structure may have had an "ethnic" basis, with different language-groups vying for control of new trade routes. Political hegemony moved east and north, centering on the Yucatán, where Uxmal and subsequently Mexican-influenced Chichén Itzá, Mayapan, and Cozumel became dominant trading towns whose well-equipped canoes plied the Caribbean coast. Modern archaeologists are generally more cautious now than they were thirty years ago about attributing these developments to single causes like cataclysmic ecological degradation or disease. They prefer the view that several interrelated social and historic factors such as political geography, foreign invasions, dynastic squabbles, collapse of trade routes, poor food distribution, and dissension among poor and marginal sectors combined to create these changes. Nonetheless, recent paleoecological data suggest that the abandonment of the big lowland cities coincided with an extremely dry period. Although the population declined, the Petén and lowland Belize were not suddenly deprived of people. In some areas—around the Petén lakes, for example—life carried on. It was in this region, presently less farmed, less populated, and more forested than in earlier centuries, that a multicultural Maya group known as the Itzá held out against the Spanish from their redoubt at Tayasal; they were finally destroyed in A.D. 1697.

The Southern Cultures

To the south, beyond the zone of influence of the great Maya towns, the development of sedentary agricultural societies followed a different trajectory. Changes in social organization took place later and more slowly than in the northern sector. There is no clear evidence of village planning, domestic or public architecture, or an upper (chiefly) class until A.D. 300. Tronadora, on the shores of man-made Lake Arenal, is the earliest pottery-using village in Costa Rica. Circular houses lie over a Preceramic occupation. The pottery, perhaps made as early as 1500 B.C., contrasts red-painted, incised, and stamped zones in a pleasing manner. It is more similar to pottery of contemporary Soconusco villages than to Panama's more ordinary Monagrillo style and shows affinities with the pottery of Ometepe Island in Lake Nicaragua, a newly discovered site near Managua, and La Montaña, near the well-known architectural site of Guayabo (Costa Rica) (see fig. 6–10D, E).

Jadeite and Gold

Another material that unifies northern Costa Rica and southern Nicaragua culturally is jadeite, which was sawn, drilled, and incised into exquisite bird and mammal figurines, often shaped like axes (see pl. 10D). Most of the raw material came from the Motagua Valley (Guatemala). This may explain why many pieces were reworked and why some, like the elongated pendant found in an important man's grave at Talamanca de Tibás, were possibly heirlooms. The pendant's Olmec- or Izapan-like design, which depicts a human hand holding a jaguarlike animal, seems out of phase with associated pottery, which dates to approximately A.D. 100–500 (see pl. 9D).

In Panama, where jadeite ornaments were not used, wealth and hierarchy between about A.D. 400 and 600 were symbolized by beautiful pendants carved in animal forms out of *Spondylus* shells (see pl. 11B). These large marine bivalves are found around coral reefs and rocky islands in water deep enough to require diving for their extraction.

After A.D. 600, the importance of jadeite and *Spondylus* shell declined as goldwork became the commodity that appealed most to southern chiefs for showing off their power and wealth. The oldest gold objects known from Panama are technologically very advanced. Their iconography suggests that metallurgy was introduced from the Sinú and Quimbaya regions of northern Colombia (see pl. 11A). But the medium was quickly adapted to isthmian intellectual requirements. Chieftains and warriors laid to rest at the famous site of Sitio Conte (A.D. 700–1000) were buried with dozens of gold pieces: hammered and embossed helmets, pectorals, and cuffs; human and animal figurines; beads, pendants, and ear spools. The animals they depict, often with human stances and faces, are also found painted on the brightly colored polychrome pottery, which is quite unlike contemporary styles from northern South America or elsewhere in Central America (see pl. 11E).

The exchange of gold jewelry became perhaps the major commercial activity of Panamanian and Costa Rican chiefdoms. Gold was rarely mined from veins in rocks; rather, it was extracted by washing alluvial gravels. Raw gold was often most abundant in remote, thinly populated regions. It was sent to large villages in exchange for products that were not locally available, such as colored cotton blankets, salt, salt fish, and hunting dogs. Cori, a chieftain who lived near modern Panama City, manufactured cast gold objects with imported nuggets. Probably the Nahua-speaking traders seen by the Spanish along the Caribbean coast of Panama and Costa Rica went there to acquire cast figurines and bells, of which several were found in the famous *cenote,* or well of sacrifice, at the Yucatán city of Chichén-Itzá. Spanish Franciscan friars on the scene wrote that gold figurines were still being cast and buried with the dead in Talamanca (Costa Rica) in the eighteenth century.

Sculpture and Architecture

The staggering wealth of Panamanian and Costa Rican chieftains was not accompanied by architectural symbols of power as ostentatious as those of the Maya and Nahua worlds. The haunting ceramic head-effigy from Miraflores in eastern Panama is perhaps a portrait of a local chieftain buried in an underground rock-cut vault (see pl. 9F). In this part of Panama, Spanish soldiers describe large wooden buildings in which the dried and smoked bodies of ancestors were preserved, some-times with gold face masks. Christopher Columbus stumbled upon sim-ilar buildings in Caribbean Costa Rica (Puerto Limón); here, flat stone slabs sculpted with animal figures were set up alongside the graves of chieftains (see pl. 11C).

Stone sculpture, in fact, is the hallmark of "head villages" of chief-doms in southern Central America. At El Caño (A.D. 500–1520), in cen-tral Panama, a ceremonial precinct demarcated by columnar basalt columns contained stone pillars surmounted by realistic carvings in human and animal form. One male figure wears a golden frog dangling from a cotton cord. Gold figurines also adorn the chests of stern war-riors from the highland site of Barriles (Gran Chiriquí), who wear coni-cal hats (probably made out of bark cloth) and sit atop the shoulders of mournful captives. Huge stone tables for grinding corn are festooned with human heads. Male and female figures that form the legs have ex-aggerated genitalia. This implies rituals involved with human agricul-tural fertility (fig. 6–17).

Barriles is within sight of the Barú volcano, which erupted about A.D. 600, covering many villages with ash and tephra. Houses were cir-cular and made of poles and palm thatch. Native peoples who once lived in Chiriquí and Bocas del Toro—perhaps ancestors of the extinct Dorasques and Chánguenas—moved into these wet, cool highlands sometime after 500 B.C.. They planted maize, kidney beans, sweet pota-toes, and wine palms. The eruption convinced many of them to leave the area for the Caribbean coast, where perennially humid conditions and broken terrain had curtailed colonization. Here, they reverted to living in hamlets alongside cultivated plots in a zone that was still largely covered with rain forest. Gardens around the carefully exca-vated Cerro Brujo site (A.D. 600–900) attracted large numbers of agoutis and pacas (brown rabbit-sized rodents), armadillos, and peccaries, which hunters would ambush while they browsed on manioc, native yams, and palm fruits.

Highland sites have such acid soils that no bone is preserved, but in lowland Parita Bay, Panama, oval and round houses occupied around A.D. 100–1500 contain thousands of animal parts in their refuse heaps, the product of many centuries' meals. As in the Maya world, white-tailed deer abounded. But only riparian forest and some hilltop woods remained undisturbed by the heavily maize-biased agriculture. Mon-keys, tapir, and curassows (terrestrial, tropical turkeylike birds) were never brought to the dinner table; agoutis and peccaries rarely. The Spanish tell us that in the chiefdom of Antatará, near Parita Bay,

6–17. Stone sculpture from Barriles, Panama: *(top):* seated man with a conical bark cloth hat sits on the shoulders of a captive. He is wearing a gold figurine around his neck and holds two human heads in his hands; *(bottom):* a large grinding stone adorned with human heads, probably post A.D. 600. From Doris Z. Stone, *Precolumbian Man Finds Central America* (Harvard University, Cambridge, 1972), figures on pages 102, 108.

Panama, the eating of deer meat was taboo at certain times. So perhaps the copious remains of painted terrapins, mud turtles, small birds like anis, and even the poisonous marine toad represent lower-class food-stuffs, consumed during periods when deer was forbidden.

The Precolumbian culture that made the Barriles statues extended westward into modern Costa Rica as far as San Isidro. Archaeologists call this region Gran Chiriquí. In later times (1100–1560), a striking un-painted terracota ware, named Bisquit, was produced here. It is charac-terized by realistic modeled animals like tapirs (see pl. 10H). A large, wealthy population lived in the productive Terraba Valley in Costa Rica, where houses were larger and more sturdy than in Panama. They have circular bases made out of river stones and are connected with stone causeways. Each one was probably inhabited by an extended fam-ily. Spanish soldiers in the 1560s report that the round houses of Coctú—located between the Térraba River and the present-day Costa Rica–Panama border—sheltered nearly a hundred inhabitants. Enor-mous balls of volcanic rock found in alignment at sites in the Diquís delta were once thought to have been shaped naturally during massive eruptions. Archaeologists now attribute their roundness to human hands. It remains to be seen whether they were used for measuring ce-lestial events (see fig. 6–3, Diquís insert). Some cemeteries consist of hundreds of stone-lined graves (see fig. 6–8A).

Residential architecture in the Central Valley of Costa Rica, where the old and new capitals of Cartago and San José stand today, also in-corporates cobble foundations and pavements. At some sites, round and rectangular structures are found (see fig. 6–3, La Fábrica insert).

The acme of Costa Rican architecture is found in the very humid, steep-sided valleys that drain into the central Caribbean coast. Here, a handful of sites, of which the best known are Guayabo de Turrialba, La Fábrica, and Las Mercedes, contain circular, stone-faced mounds (prob-ably foundations for elite residences), cobble-paved causeways, stair-ways, plazas, and aqueducts (see fig. 6–3, Guayabo insert). These sites bear a striking resemblance to Tairona villages in the Sierra Nevada de Santa Marta in Colombia. Ceramics from this part of Costa Rica are renowned for vessels that have large tripod feet with masterfully mod-eled forest reptiles, mammals, and birds (see pl. 10G).

The Prevalence of War

Spanish military commanders found the native peoples of the highlands of Costa Rica and Panama tough nuts to crack. Spanish cavalry with firearms and pike- and sword-wielding infantry were better equipped to fight massed armies in open country than mobile bands armed with palm-wood spears, bows and arrows, and blowguns in forested country. Historians are reluctant to attribute the Spanish withdrawal from trading and mining sites occupied in the sixteenth century, for example, the Valle del Guaymí (Panama) and Talamanca (Costa Rica), to the military prowess of these mountain polities. Yet they were always bickering about land, hunting rights, stolen women, and murdered captives, and their bellicose lifeways are amply reflected in their stone sculptures of human figures with elegant tattoos, padded belts, and helmets. Often they carry axes and trophy heads (see pl. 11D, fig. 6–17).

The prevalence of war is perhaps reflected in the iconographic preeminence of the jaguar, which was the cult animal par excellence over much of Costa Rica and western Panama. Thousands of stone tables, stools, and columns with jaguar features have been ripped out of graves by unscrupulous looters. They evoke chieftains bedecked with macaw- and quetzal-feathered headdresses, gold necklaces, and earspools holding counsel while seated upon jaguar stools, symbols of strength and cunning (see pl. 10E).

A profound realism also characterizes the arts and crafts of the Gran Nicoya region of southwest Nicaragua and northwest Costa Rica. Between 500 B.C. and the Nahua incursions a millennium or so later, a thriving local culture produced a beautiful pottery called zoned bichrome because of its shining red- and black-painted areas separated by incised lines (see pl. 10F). Modeled human and animal forms abound; among the animal subjects are amphibians, birds, tapirs, and armadillos. Polished stone mace heads (see pl. 10C), which must have dealt horrible blows, present a combination of realism and stylization that is comparable in excellence to Inuit handicrafts.

With the arrival of Nahua-speaking peoples from the north about A.D.700 came a change in art styles. Polychrome painting now predominates. Designs include depictions of such northern culture gods as the Nahua-speakers' rain-god, Tlaloc (see pl. 11E). Rampant head-taking is apparent in burials of warriors surrounded by skulls.

The Impact of Conquest

On the eve of Spanish conquest and colonization, Central America was a well-populated section of the New World tropics. In areas of seasonal climate where the natural vegetation had been cut and burnt for millennia, the landscape was a variegated patchwork of native settlements, fields, and gardens intermingled with grassy tracts, second growth, and wooded river valleys and hills. Tall, mature forests remained in many mountainous and humid areas, especially on the lower Caribbean slopes south of the Gulf of Honduras. Some had been altered little by human activities; others supported hamlets of farmers, fishers, hunters, and traders scattered along coasts and up rivers.

The Spanish chronicler Gonzalo Fernández de Oviedo visited northern Costa Rica and the culturally related lake district of Nicaragua in the late 1520s and describes precisely a well-populated landscape with large towns, markets, and fairs.

In important details, however, this very "anthropogenic" environment would have looked quite different from today's. Nestled amid tall sapote and avocado trees and palm groves, towns and villages would have seemed smaller than they really were to a distant observer and more verdant than most modern settlements. Ungrazed by horses, mules, or cows, grasslands would have had an unfamiliar look. Fields (especially Maya fields) were neatly furrowed or mounded; but, in the absence of metal axes, forests were probably not cleared as completely as a machete-wielding campesino would clear them today. Mixed and tangled vegetation in and around abandoned fields offered shelter for surprisingly large herds of white-tailed deer and, where there were few people, collared peccaries.

Culturally, too, the isthmus was heterogeneous. Columbus was startled by the myriad customs and languages of the Caribbean coast of Central America in 1502–03: "The villages have each a different language, and it is so much so that they do not understand one another any more than we understand the Arabs." As we have seen, some of these languages represented an in situ diversification of ancestral Central American tongues; others accompanied dislodged peoples who entered the isthmus from the north.

Unlike Mexico and Peru, contact-period Central America was not ruled by an imperial power. Rather, city-states, ethnic and mercantile confederations, and, in the southern sector, hundreds of small chiefdoms vied for power and wealth, clashing continually in strident wars and raids. In spite of these conflicts, however, trade and barter flourished. Merchandise, transported on porters' backs along well-trodden trails or in huge dugout canoes, was exchanged in busy markets and at periodic feasts and ceremonies.

The Spanish Conquest (1502–42)

The Spanish conquest and colonization radically and rapidly transformed Central America. In two converging waves and in only forty years (from 1502 to 1542), single-minded captains avid for instant wealth and the social recognition it conferred carried Spanish arms from one end of the isthmus to the other (pl. 12). After Columbus's unsuccessful attempt to settle the windswept coast of Veraguas (Panama), Santa María (1510–24) and Acla (1515–48) became the bridgeheads of Spanish raiding along the San Blas coast, the Gulf of Urabá, and across the central *cordillera* to the Pacific coast, "discovered" by Vasco Núñez de Balboa in 1513. Noticing that larger and more nucleated populations existed in the Pacific coastal savannas than on the wetter Atlantic slopes, Pedrarias Dávila, governor of the Spanish province of Castilla del Oro, founded Old Panama in 1519 and Natá in 1522, driving native resistance into the Atrato basin of Colombia and the interior mountains. Exploiting Darién and Pearl Island forests to build ships, he dispatched other soldiers (pl. 12)—Juan Ponce de León, Gasparte Espinosa, and Gil González Dávila—by sea and land up the Pacific coast of Central America to the Gulf of Fonseca. News of populations even larger and culturally more advanced than those of eastern Panama convinced Pedrarias to take his rapacious habits to lacustrine Nicaragua and Nicoya, where the Spanish had established a bridgehead at Bruselas as early as 1524. Displacing the more humane Fernando Hernández de Córdoba, he converted Nicaragua into his personal fiefdom.

In 1524 Pedro de Alvarado, a lieutenant of Hernán Cortez, conqueror of Montezuma's Mexico, brought a force southward through Chiapas into Guatemala and El Salvador (pl. 12). Exploiting the intergroup rivalry that characterized the Mexicanized highland Maya city-states, he stormed the Quiché fortress at Iximché with Cakchiquel allies from Utatlán and then turned on the Nahua-speaking Pipil. Slave raiders from the Spanish West Indies had been making destructive raids into the Gulf of Honduras for a decade before Cortez himself, making a tortuous expedition from Mexico, was forced to snuff out internecine squabbling.

Ravaged by smallpox, measles, pulmonary plague, and other Old World diseases to which they were not immune, racked by internal dissension, and mesmerized by the strangeness and histrionic violence of the invading culture, native communities offered little effective resistance to this double-headed invasion. Populations plummeted. In Panama, the once-populous confederation the Spanish called Cueva were obliterated within a few decades. The number of estimated native tributaries in early Spanish Nicaragua declined from 600,000 in 1520 to 6000 in 1560–70.

Understandably, the Precolumbian subsistence and exchange economy collapsed. Abandoned towns and villages vanished beneath the forests that quickly invaded agricultural landscapes. The lot of most surviving populations, whether they resisted or not, was enslavement, either for labor gangs in the gold and silver mines of Honduras (Olancho) and Nicaragua (Nueva Segovia) or as so-called grants to individual Spaniards for domestic and agricultural services, ostensibly in return for catechization.

At first, the native labor supply seemed infinite, so little effort was made to mitigate suffering and death. Tens of thousands of native slaves were exported to richer parts of the rapidly expanding Spanish empire, particularly during the pacification of Inca Perú in the 1530s. Loss of cultural identity was hastened by the Spanish soldiers' practice of cohabiting with several native women. A Hispanicized mestizo (mixed blood) population proliferated rapidly around the scattered, thinly populated Spanish settlements. In the face of these events, it is surprising that native Central Americans survived at all. But survive they did, in many forms and guises and by dint of very different strategies (see chapter 8).

Church and Labor Grants (Encomiendas)

The depredations of the first wave of Central American colonists shocked some educated sectors in the mother country. Debates about the morality of indiscriminate killing in the name of Christ had gone on between such learned churchmen as Bartolomé de Las Casas, and Juan Ginés de Sepúlveda ever since the first Spanish galleons arrived in the Caribbean. Ultimately, the pronative lobby prevailed, and legal measures were taken to protect the natives. The most comprehensive were the famous *New Laws,* promulgated by Charles V in 1542. In theory, the Indians, as they were called, would henceforth be free vassals of the emperor. Forced labor, Spanish polygamy with native women, work in the pearl fisheries, and slavery (unless justified by military confrontation) were abolished. The labor grant (*encomienda*) system (forced labor of Indians for the original conquistadors in return for being Christianized) was abrogated. On the death of a grantee (*encomendero*), native servants were to become charges of royal authorities, the Church, or religious orders.

In practice, the success of the *New Laws* and subsequent humanitarian legislation was linked to the honesty, energy, and political acumen of individual Crown and Church officials. Spanish colonists often turned a blind eye to these irksome legalities. The labor grant system continued to function de facto. Settlers saw no reason to stop exacting tribute in labor and produce from satellite native communities.

The natives' responses to the new social relations that were forced upon them varied greatly throughout the isthmus. Partly because they were more densely populated in Precolumbian times and partly because their societies were accustomed to structured hierarchies, the Indian towns of the ancient Maya and Nahua spheres retained their native language and customs much longer than did speakers of Misumalpan and Chibchan tongues to the south. In Chiapas, Quetzaltenango, Huehuetenango, and Verapaz, native communities tried to preserve their Indian culture, although mixed with some Spanish elements. On the other hand, most Indian towns in Nicaragua and Panama had lost their native languages by the beginning of the seventeenth century; even so they continued to be treated as distinct cultural and political entities throughout the colonial period.

Embittered by economic depression, colonists did their best to provoke new conquests in remote regions in order to legitimate and replenish their labor supply. This attitude characterized the conquest of highland Costa Rica, which did not begin in earnest until the 1570s. It was to continue, on and off, throughout most of the colonial period and was often the *modus causandi* of native rebellion.

The organization of native populations that remained under the aegis of the Crown was increasingly entrusted to Dominican, Franciscan, and Mercedarian friars. They supervised the congregation of "reduced" natives and made specific evangelical journeys to "unreduced" communities to persuade them to adopt the Christian faith and Spanish ways. Generally, spiritual conquest was more beneficial for conquerors and conquered than conquest by force of arms. One has to admire the sense of purpose and indomitable will of prelates such as Fray Adrián de Santo Tomás, a Franciscan from Flanders who, during the first half of the seventeenth century, worked tirelessly to convince bands of warlike Guaymí and Kuna to leave their forested redoubts and follow the way of the Lord in savanna villages. In the long run, however, the combined forces of religion and tributary work deprived most of these reduced peoples of their remnant Indianness.

Native Resistance and Survival

Spanish settlement of Central America was, with a few exceptions, strongly concentrated on the Pacific side of the isthmus. Well over half of the isthmus remained remote from Spanish settlement and economic activities during the colonial period. This inaccessible, mostly forested territory was home to native polities that had survived the dark decades of conquest, some virtually unchanged since contact, others cultural and genetic hybrids. Some Indian groups, for example, the Itzá, Pech (Paya), Coto, Dorasque, Chánguena, Coclé, and Huétar, were destroyed or subjugated before the colonial period was over. Their remnants were forced into towns, where they were Christianized and Hispanicized and ultimately amalgamated into the poor masses of empire. Others, such as the Lacandones of Guatemala and Chiapas, the Miskito of Caribbean Honduras and Nicaragua, the Bribri and Cabécar of Costa Rica, and the Kuna and Guaymí of Panama, carved out territories that they continued to defend resolutely against Spaniard and native enemy alike into modern times.

The Kuna of eastern Panama, first mentioned by name in Spanish documents at the end of the sixteenth century, were visited in the 1640s by Fray Adrián de Santo Tomás. By this time, their settlements were widely dispersed around the shores of the Gulf of Urabá (Colombia) and along the rivers of Darién. Taking advantage of the spread of post-conquest forests, they launched effective and destructive raids deep into Spanish territory. Although the Crown responded by trying to Christianize the population, promoting civil discord, and building forts, the Kuna retained their independence, signing a famous treaty with the Spanish Crown in 1765.

Armed resistance to the Spanish began in central and western Panama soon after Natá was founded in 1522. The abandonment of the rich Veraguan mine of Turlurlí in 1589 is attributed by historians to rapid exhaustion of veins; but constant native attacks from unconquered territory surely influenced the decision to retreat. From this date until republican times, the entire Caribbean coast between Guaymí territory and central Costa Rica defied Spanish colonization. Major revolts in 1610 and 1709 by so-called Talamancans (Bribri and Cabécar) terminated the missionization process and retarded acculturation until the nineteenth century.

In the eighteenth century, Talamancans and Guaymí came increasingly under pressure from the highly mobile Miskito, who carved out a wide sphere of influence down the Caribbean coast at the expense of both Spanish colonists and neighboring native groups. Like the Kuna, the Miskito exploited European rivalries intelligently. Acquiring firearms and watercraft from the English in the Caribbean, they became a formidable fighting force. Their mobile war parties raided deep into Spanish territory, disrupting trade and causing loss of life. Their pacification in the nineteenth century was due as much to changes in British colonial policy as to Spanish resistance. They last raided Panama (Santa Fé) in 1805, only ten years before the Battle of Waterloo.

CHAPTER SEVEN

Spanish Rule, Independence, and the Modern Colonization Frontiers

STANLEY HECKADON-MORENO

The Effects of Geography and Climate

Central America is a land dominated by mountains: they occupy about three quarters of its surface area. Pacific and Atlantic coastal plains of varying widths flank the central mountains, although they are narrow by comparison (pl. 13A). Until recently Spanish colonization was strongly centered on the Pacific coastal zone and the valleys and plateaus of the Pacific slope of the central mountains. The mountainous topography hindered the development of the Spanish administration because it made communications difficult and costly between and within the regions. After independence, it was an obstacle to the consolidation of national markets. Until the recent past it was, in many cases, easier for coastal towns to communicate with foreign ports than with their own hinterlands. In the political sphere, the physical isolation of the regions abetted political fragmentation and instability.

The isthmus is dominated by an active volcanic chain, and it is no accident that most national flags, coats of arms, and symbols have volcanoes as a prominent motif. Both volcanoes and earthquakes have marked the land and the lives of men. Volcanoes have been a blessing for the fertile soils they have generated and a curse for the destruction that they have regularly caused. The ancient Indian civilizations considered volcanic activity, or the lack of it, an omen of events that would deeply affect their lives and societies. When volcanoes were quiet, the gods were angry and terrible events would ensue. According to the annals of the Cakchiquel Mayas, the secret of fire was stolen from the fiery entrails of the Gagxanul volcano.

Most capital cities in the isthmus have suffered devastating earthquakes with major loss of life and property. Guatemala City has been punished the most, having been severely damaged on nineteen occasions, the last in 1976. Managua, Nicaragua, still shows the scars of the quake of December 1972, and in 1986 San Salvador, the capital of El Salvador, was shattered (fig. 7–1). Thousands died, and economic damages ran into hundreds of millions of dollars.

Rainfall is concentrated mostly on the Atlantic coastal zone of the isthmus, where it has created spectacular rivers, many with lovely indigenous names: Motagua, Ulúa, Rama, Reventazón, Sixaola, Changuinola, and Chagres. In this region there is much less seasonal variation in rainfall than on the Pacific coastal zone. Peasants from the Pacific side joke that on the Caribbean it rains thirteen months a year. The rivers on the Pacific side are shorter, carry smaller volumes of water, and flow more irregularly; about 90 percent of their discharge runs into the sea during the rainy season. As this region has become densely populated there has been a growing scarcity of water, now reinforced by the destruction of watersheds.

In the Pacific and central mountainous zones, there are two well-differentiated seasons; the dry season from December to May and the

7–1. Damage caused by the earthquake of 1986 in San Salvador, El Salvador. Photograph by Stanley Heckadon-Moreno.

rainy season for the rest of the year. The alternation of these two seasons marks the rhythm of the agricultural calendar.

The Pacific coastal plains are usually narrow (see pl. 13A), but volcanic activity has deposited extensive, deep layers of ash that have produced some of the planet's richest agricultural soils. These volcanic soils, however, are easily eroded.

The broader coastal plain of the Caribbean is swampier, and the coastline is dotted by lagoons, estuaries, and lakes. The soil is less fertile, has greater acidity, and is subject to serious drainage problems. Here the best agricultural soils are found along the floodplains of the great rivers. It was in these sites that the later banana plantations would take root.

From the sixteenth century on, Spaniards have chosen to settle in the drier central highlands and along the Pacific coast. In these areas mestizos, descended from Indians, Spaniards, and Blacks, have gradually increased in numbers until today they are the predominant ethnic type of El Salvador, Honduras, Nicaragua, and Panama. Historically, the Spaniards left the wet forests and swamps of the Caribbean to the Indian and Black communities. On the Caribbean coast, the cultural and ethnic mixtures of Indians and Blacks were augmented by traders and pirates from England, France, and Holland, giving rise to diverse new groups, so that the Atlantic side of the isthmus is now a distinctly different social environment from that in the central mountainous zone and the Pacific coast. Culturally and linguistically, people on the Caribbean coast evolved in easy contact with English-speaking peoples, especially from Jamaica and other Antillean islands. To this day, it is customary for Caribbean coastal peoples in some countries to refer to the mestizos from the Pacific side as Spaniards.

To the natural factors described above, one needs to add the strategic importance of Central America's geographical position, which has contributed significantly to its socioeconomic evolution. The narrow isthmus has been the most important route for interoceanic transit and communication between the East and West, and this "geographic destiny" is most evident in Nicaragua and Panama. Conflicts between the world powers for control over the interoceanic route, including the neighboring Caribbean islands and the most important straits that give access to the Central American isthmus, began as long ago as the sixteenth century.

Central America under Spain (1500–1821)

From Cuba and Hispaniola, the Spaniards organized the conquest of the American mainland. From the islands of the Antilles, Spanish settlement moved onto the mainland in two principal regions: Mexico (New Spain) and Panama, or Tierra Firme. The first wave of the conquest was two pronged (see pl. 12 and chapter 6). The first, led by Pedro de Alvarado, pushed southward from Mexico, following the Aztec trading routes toward Guatemala and Nicaragua. Alvarado, a captain of Hernán Cortez, conquered the upper section of Central America, including Chiapas, Guatemala, El Salvador, and Honduras. The second advance came from Panama, Pedrarias Dávila leading a force through Costa Rica to reach as far as Nicaragua.

The discovery, conquest, and establishment of administrative order over the vast territory of Central America was an extraordinary feat accomplished in roughly thirty years by small bands of conquistadors. Rapacious yet courageous, these adventurers had an unquenchable thirst for wealth and used a few technical advantages like the horse, guns and iron, and astute alliances with rival indigenous groups to conquer hundreds of Indian chiefdoms and kingdoms. In the process, they won millions of vassals for the king of Spain and souls for the Catholic Church. Under Spanish rule, the region had for the first time a common language and religion.

The arrival of the Spaniards triggered a major exchange of plants and animals between the Old and the New Worlds. They made known to Europe cacao, tobacco, beans, pineapple, and peanuts. In turn, they brought to the Americas rice, coffee, citrus, and sugar cane as well as cattle, horses, and other livestock. The only domesticated animals Precolumbian Central America knew were dogs, turkeys, and stingless bees. When the conquest decimated the Indians of the savannas and grasslands of Honduras, Nicaragua, and eastern and central Panama, these ungrazed lands provided excellent fodder for herds of Old World cattle (see pl. 13B).

European livestock entered Central America via Panama in 1521. Cattle did better than pigs, and by the later 1520s the newly colonized lands around the city of Natá in central Panama had large herds. So many ships bound for Peru were provisioned at Old Panama that the demand for salt meat, hides, and tallow outstripped the local supply. This demand stimulated the growth of livestock rearing in Nicaragua and the Costa Rican province of Guanacaste, which is climatically similar to the Pacific coast of central Panama. Early Spanish colonists dreamed that cattle haciendas would bring them increased wealth and status, but this rarely happened. Many large properties were carved out of the ancient savannas, but most became sleepy, if self-sufficient, domains of a few Creole families, who often found smuggling and public office more profitable ways of accumulating wealth.

The Spaniards also introduced the structure of the Mediterranean city, wheeled carts, and the foundations of the social and religious institutions that endure to this day.

The *Audiencias* of Guatemala and Panama

The colonial heritage is critical to understanding the formation of the national states after independence. For more than 300 years (1500–1821), Central America was an administrative unit of the Spanish Empire. Spain divided its New World domains into four vice royalties: Mexico, New Granada (Colombia), Peru, and the Plate River. In colonial times, Central America was divided into two jurisdictions called *audiencias*, one in Guatemala and the other in Panama.

The audiencia of Guatemala extended from the present Mexican state of Chiapas to Costa Rica. The scarcity of valuable minerals made the audiencia a poor colony, except for Guatemala itself, and it was ruled and subsidized monetarily by the powerful vice royalty of Mexico. Initially a separate audiencia, Panama was later placed under the vice royalty of Peru and eventually, as its strategic importance as the main trading route of the Americas grew and as raids by pirates from rival European powers increased, under the vice royalty of New Granada.

Panama became a centrally important audiencia once the riches of the Incas began to flow across the isthmus. Some 60 percent of all precious metals that entered Spain from the New World passed between Old Panama and the Caribbean ports of Nombre de Dios and, after 1579, Portobelo (see pl. 13B). The great fairs that accompanied the arrival and departure of the galleons and the activities generated by the transisthmian route created a prosperous mercantile economy in constant demand of services. Many of these were provided by imported African slaves—by 1575 there were 8630 Black slaves in Panama. In spite of multiple interruptions, transisthmian commerce has remained the hub of Panama's economy until the present day.

To administer and exploit the newly conquered territories, cities were established (see pl. 13B): Panama in 1519; Natá, also in Panama, in 1522; León and Granada in Nicaragua and Santiago de Guatemala

(Guatemala City) in 1524; San Salvador in El Salvador and Trujillo in Honduras in 1528; Choluteca, also in Honduras, in 1534; San Pedro Sula and Gracias a Dios in Honduras in 1536; and Comayagua, Honduras, in 1537. Each city would in turn become a province. By the mid–sixteenth century, the most important nuclei of Spanish political power in Central America were Guatemala City, León, and Panama City. Later, in the seventeenth and eighteenth centuries, other cities rose to challenge the dominance of these older centers and contributed to the fragmentation of political power that was to take place after independence from Spain.

Guatemala's official, full designation was Audiencia y Cancillería Real de Santiago de Guatemala en la Nueva España (The audience and royal chancellorship of Guatemala in the New Spain). The kingdom of Guatemala, as it was also known, included the provinces of Guatemala, El Salvador, Honduras, Nicaragua, and Costa Rica and the modern Mexican state of Chiapas. The provinces were divided into *alcaldías,* or municipalities. The highlands of Costa Rica were not colonized by the Spanish until the 1570s, and expeditions sent out from the capital Cartago came into conflict with adventurers pushing north from the audiencia of Panama. Both groups suffered from fierce native resistance.

In the audiencias, people did not elect their representatives. All critical colonial functionaries, from the captain general to the provincial governors and the *alcaldes,* or mayors, of the most important municipalities, were named by the king. The men holding these posts in the Indies were customarily native-born Spaniards selected, through favor in the Spanish court, to buy the positions and titles.

The Growth of Provincial Power and Fiefdoms

Many political scientists have characterized Central American political institutions and governments as authoritarian, with centralized power concentrated in few hands, frequently that of political bosses or military strongmen known as *caudillos.* Within this culture, to dissent was to be labeled seditious and then subversive. The origins of this authoritarian, politically intolerant tradition hark back to the protracted colonial period. The colonization and settlement of Central America occurred during the time when royal despotism, in alliance with religious intolerance, expelled Moors and Jews from Spain. The civil and political liberties gained by the peoples of the Iberian Peninsula during the 800-year struggle against Islam were lost. Thereafter, all government functions were concentrated under the figure of the king and his sycophantic court. The deadening hand of excessive centralization and of an absence of governmental accountability took hold early in the life of the new colonies. A few monarchs were talented and dynamic but most were inept.

Political despotism was aggravated by religious intolerance. The instrument in the push for religious unity and the rooting out of sin and heresy was the Holy Tribunal, or Inquisition. The first tribunal in the New World was instituted in Lima in 1569. The kingdom of Guatemala fell under the holy office of Mexico, created in 1570. Panama owed allegiance to Lima and, after 1610, to the tribunal at Cartagena. In its rooting out of heresy, the tools of the Inquisition became imprisonment, torture, and confiscation of properties. Religious intolerance, by stifling inventiveness and new ideas, promoted economic and social stagnation.

The physical isolation of regions and towns facilitated the institutionalization of fiefdoms, or *caciquismo,* a form of local government based on family ties. In each province, political power tended to concentrate in the hands of a reduced number of local families, usually intertwined by marriage ties.

A good example of these centripetal forces at play was the political struggle that developed between the provincial municipalities and Guatemala City. Guatemala City was the official seat of royal power, entrusted with collecting taxes and controlling the internal and external trade of the kingdom. By the prevailing mercantilist laws, the colonies existed for the sole benefit of the metropolis, the result being that the merchants of Guatemala City monopolized most lucrative activities like the marketing of indigo, the buying and selling of cattle and precious metals, and foreign trade. Guatemala City also controlled the appointment of the new breed of colonial administrators called criollos, or Spaniards born in America. The rule by Guatemala City was fiercely contested by the elites of other economic centers emerging in the provinces. The privileges of the metropolis generated a hostile climate between the merchants of Guatemala City and, for example, the indigo producers of El Salvador and the livestock rearers of Honduras, Nicaragua, and Costa Rica. Indeed, so aggrieved were Costa Rican producers that they petitioned the Crown to be annexed to the audiencia of Panama.

Gradually, the provinces drew apart from the center of power. Given the physical distances, the poor communications, and perennial fiscal poverty, Guatemala City was unable to control the emerging local power groups, who eventually came to enjoy relative autonomy. Owing to conflicts and rivalries, the provinces also grew apart, as did the various cities within them.

Independence and the Emergence of the Republics

The independence of the kingdom and provinces of Guatemala from Spain and the emergence of the separate countries of Central America had internal and external causes. By the end of the colonial period, in the last half of the eighteenth century, there was a widespread economic crisis due to the collapse of the most important export crop, indigo (see pl. 13B). Economic impoverishment was aggravated by events in Europe, especially the French Revolution with its dissemination of republican and liberal ideas and Napoleon's invasion of Spain and dismissal of the Spanish monarchy. These events accelerated the breakdown of the Spanish Empire in America.

For 300 years the Spanish Crown had held the American colonies together against the centrifugal forces of regionalism. When, suddenly in 1808, Spain and the colonies found themselves without a king, the peoples of Central America continued to consider themselves subjects of Spain. Armed revolt started first in the most powerful vice royalties. In 1821, Gen. Agustín de Iturbide declared Mexico's independence from Spain and demanded that Guatemala join the new federation. That year, the captain general of Guatemala and the powerful merchants of Guatemala City, without consulting other provinces, declared Central America independent of Spain and annexed to Mexico. Thus, from 1821 to 1823 the provinces of Central America were part of Mexico. Panama, after declaring its independence in 1821, joined the Colombian Confederation, which included Venezuela, Ecuador, and Peru. It would remain a part of Colombia until 1903.

The second independence of Central America, this time from Mexico, began in 1823 with the formation of the United Provinces of the Center of America, which lasted until 1840. This period was chaotic, marked by constant conflicts between liberals and conservatives. In the context of Central American history, a liberal advocated free trade with other countries, a federal government that included provincial representation, and separation of the state from the Catholic Church. Conservatives upheld protection of local markets, a central national government, and Catholicism as the official religion of the state.

From 1823 to 1840, conflict spread between the Catholic Church and liberal rulers and between provinces and the capital Guatemala City; and rivalries deepened into strife among cities competing for the control of the same region: Comayagua and Tegucigalpa in Honduras, San Vicente and San Salvador in El Salvador, Granada and León in Nicaragua, and Cartago and San José in Costa Rica. The resulting political instability led to the breakdown of the federation. The five old provinces then became sovereign states within Central America. Chiapas, however, chose to remain a part of Mexico. The split of Central American society into liberal and conservative camps then led to a protracted series of civil wars and revolutions that lasted until the early twentieth century.

The Conservative Restoration (1840–70)

From the dissolution of the federal experiment in 1840 until 1870, conservative governments ruled the now separate countries of Central America. These governments were usually led by generals. Preeminent among them was Rafael Carrera in Guatemala, who attempted to restore colonial institutions and return privileges to the church. He forged an antiliberal alliance, first internally, based on support from the Indian communities, the landed aristocracy, and the church, and then gradually abroad, with conservative rulers whom he helped put into power in El Salvador, Honduras, and Nicaragua.

This period was marked by the final crash of indigo production and by two important foreign conflicts related to the growing possibility of building an interoceanic canal. The first conflict was with Great Britain, who aimed at controlling, via the Mosquitea protectorate, the Caribbean coast of Central America. The second conflict was the National War, which occurred when the separate countries collaborated briefly to face and defeat the threat from the North American filibuster William Walker. On behalf of the interests of the Morgan Trust, Walker had invaded Nicaragua seeking to establish the rights to a waterway across Nicaragua. The National War had the unintended effect of militarizing the national societies, including the then-distant and isolated Costa Rica.

One of the most important achievements of the conservatives was the establishment of large-scale cultivation and exportation of coffee (pl. 14A). By the end of the conservative era in 1870, however, the Central American States were more different from one another and less compatible than they had been in 1840. The dissolution of the Federal Republic, civil wars, and political instability had disintegrated the former economic unity of the region. From now on, each of the small new nations became distrustful of the others.

From 1870 until the Great Depression of 1930, the political pendulum shifted, and countries came to be ruled by liberal governments. This was a dynamic period of substantial reforms and modernization brought about by the increased development of the coffee industry and, later, the production of bananas (see pl. 14A). The social and economic changes brought about by the cultivation of coffee were felt most profoundly in Guatemala, El Salvador, and Costa Rica. In Honduras and Nicaragua, sustained political turmoil retarded cultivation, and the main economic export activity remained silver and gold mining by foreign investors.

This period saw the emergence of a new coffee oligarchy, which came to control the economy and the politics. Export crops such as coffee permitted a new fiscal structure whose revenues were invested by the state in infrastructure: schools, hospitals, roads, railroads, and electricity.

By the end of the nineteenth century, small industries had arisen in the towns along with an urban working class. But, to facilitate the expansion of coffee farming, governments expropriated and sold to more powerful interests the common lands of the Indian and peasant communities and those of the church. These lands had been held through tenure arrangements made during the colonial period. Now, for the first time, registries for land property were established and foreign investment was attracted with generous land and tax concessions. The Central American governments also stimulated foreign immigration, particularly from Europe. By the century's end, many Italians and Belgians had immigrated to Costa Rica and Germans to El Salvador and Guatemala. Most became linked to the production, processing, and exportation of coffee.

The strong rulers who executed the liberal revolution were also mostly generals: Justo Rufino Barrios and Manuel Estrada Cabrera in Guatemala, Marco Aurelio Soto in Honduras, José Santos Zelaya in Nicaragua, and Tomás Guardia in Costa Rica. Distrustful of the masses and of the church, liberal dictators gradually strengthened the military to enable them to impose their agenda; later, the armies would become a major threat to the institutionalization of democracy.

The liberal modernization brought about the marginalization of many rural producers and the emergence of a class of landless peasants and an urban laboring class. Prosperity based on coffee and bananas lasted until World War I, when exports to Europe and the United States collapsed. Trade with Europe, until then the main trading partner of

Central America, diminished, and thereafter commerce became reoriented toward the United States.

Contemporary Dilemmas

The Demographic Revolution

There is a long-standing debate about the size of the indigenous population of Central America at the time of European conquest. The estimates vary from 1 to 10 million. What is certain is that upon contact with the Europeans, the native population dropped dramatically from several million to a few hundred thousand in only five decades. This demographic holocaust was due to several factors, including warfare and economic exploitation, but the major cause was disease from the Old World, against which the Indians had no resistance.

During the 300-year colonial period, the Spanish and mestizo population grew slowly, requiring about 100 years to double. After independence from Spain, the growth rate increased slightly until the twentieth century, when a demographic revolution occurred. Gradual improvements in public health, education, and general living conditions were the basis for this social transformation in which the doubling time of the population was decreased from 100 to 25 or 30 years. Central America began the century with 3 million inhabitants; by 1990 it exceeded 30 million (fig. 7–2).

This unprecedented phenomenon is having an enormous impact on natural resources and social structures throughout Central America. Furthermore, the modern populations are highly skewed, with 44 percent of the people aged less than fifteen years. Some 50 percent are concentrated in the Pacific coastal zone and about 40 percent in the central mountains, leaving only 10 percent living on the extensive Caribbean coastal plain. And in each country, the population is tending to concentrate in reduced areas, especially around the capital cities. In Panama, two-thirds of the population live near the urban centers along the interoceanic waterway between Panama City and Colón. A similar proportion of Costa Ricans reside in the Central Valley. The Hondurans are concentrated in the central corridor of Choluteca-Tegucigalpa-Comayagua, and in Guatemala on the Altiplano, or high plateau.

Rapid population growth has also resulted in the intensification of forest colonization. During the first half of the twentieth century, the agricultural frontier penetrated the last uninhabited jungles of the Pacific slope. Typical examples are the areas of La Máquina and Concep-

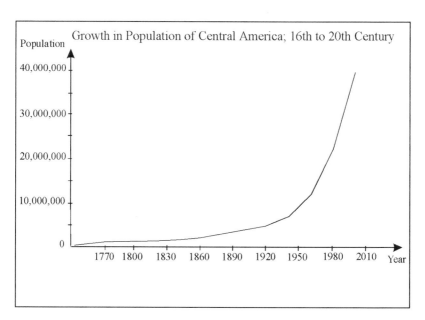

7-2. Growth in population of Central America, sixteenth to twentieth centuries. From R. L. Woodward, *Mesoamerica*, vol. 1, p. 229, 1980; CELADE, Boletín Demográfico No. 27 and 35, 1981.

ción in Guatemala, the coastal plains of El Salvador, the region of Golfito in Costa Rica, and Barú and Tonosí in Panama.

Later, after World War II, thousands of peasant families began to migrate from the densely inhabited and ecologically degraded areas of the Pacific side of the isthmus and the central mountainous zone toward the rain forest on the Caribbean side.

The Agro-Exporting Economy

Since the sixteenth century, Central America has had what historians call an agro-exporting economy, the central theme of which is the search for a crop whose exportation will generate wealth. These economies have thus been oriented toward a monoculture whose excessive dependence on volatile external markets has resulted in alternating cycles of booms and depressions.

During the colonial period, there were monocultural booms in balsam, sarsaparilla, cochineal, cocoa, and indigo. Grown on the Pacific side (see pl. 13B), these crops were each the basis for varying degrees of riches and power. Cacao fueled the first agricultural bonanza, occupying the best volcanic soils of the Pacific. Its cultivation expanded through Guatemala, El Salvador, Honduras, Nicaragua, and, to a smaller degree, Costa Rica, and its main market was Mexico. During the seventeenth and eighteenth centuries, most of the wealth came from the cultivation and extraction of vegetable indigo. This industry was to disappear in the mid–nineteenth century with the discovery in Europe of synthetic dyes.

Coffee and Bananas

Since the second half of the nineteenth century, a major social and economic revolution was brought about by the introduction of the cultivation of coffee and bananas. It is impossible to understand contemporary societies in Central America without making reference to the coffee farms and the banana plantations.

Coffee was planted on the high-quality upland volcanic soils of the Pacific, as high as 1000 meters, to be sold mostly to Europe; between 1860 and 1900, the development of coffee growing in Guatemala and El Salvador was impressive. Guatemala, which in the 1850s exported barely 15,000 quintals a year, in 1885 sold 4 million quintals (a quintal weighs 100 kilos). Bananas were initially cultivated on the fertile alluvial valleys of the broad rivers of the Atlantic coast in Guatemala, Honduras, Costa Rica, and Panama (see pl. 14A). Later, in the first decades of the twentieth century, bananas were also planted over large areas of the Pacific coast of Guatemala, Costa Rica, and Panama (see pl. 14B). The cultivation of bananas was the domain of large foreign enterprises, mostly from the United States, that supplied the North American market. Because bananas developed in sparsely inhabited, isolated forest areas, huge investments in infrastructure were necessary, and workers had to be brought in. On the Atlantic coast, thousands of mostly Black laborers from the Antilles came to the plantations; of Protestant religion and English-speaking, they greatly diversified the cultural and ethnic character of the region. In turn, large waves of mestizo peasants migrated to the plantations on the Pacific side: Salvadorians to the sector of Tiquisate, Guatemala, and Nicaraguans to Costa Rica and Panama. Throughout Central America, banana plantations were to become one of the main vehicles in the transformation of the traditional peasantry into rural wage laborers.

The cultivation and export of coffee and bananas were made possible by a transportation revolution based on the train and the steamship. Great investments were made in railroads, roads, ports, electricity, and communications to hasten the delivery of these products from the production areas to the external markets at the lowest cost. In the nineteenth century, Central American coffee destined for export was taken to ports on the Pacific coast. Steamships would deliver it to Panama, where it was transported by railroad across the isthmus to the Caribbean port of Colón for shipment to Europe and the United States.

The ways in which national societies organized themselves around these crops differed. In Costa Rica, coffee farming developed around small properties. Some social researchers have suggested that this early orientation toward economic democracy was to set the stage for the growth of Costa Rican political democracy. In Guatemala and El Salvador, liberal governments facilitated the expansion of coffee farming by expropriating the common lands of Indian and mestizo peasant communities. Thus, production became centered on a few large farms owned by a powerful and closely knit elite, the so-called coffee oligarchy, who also came to control political power. This bred economic inequality and was to produce serious political strife subsequently.

Cattle, Cotton, and Cane

After World War II, the agro-exporting economies of Central America became much more complex. Areas dedicated to cattle and new crops such as cotton and sugar cane were expanded. The most aggressive expansion was that of the cotton plantations along the southern coast of the isthmus (see pl. 14B). Between 1950 and 1970, cotton farming from Guatemala to Guanacaste in Costa Rica ensured the rapid disappearance of the last dry Pacific zone forest. In El Salvador, the construction of the coastal highway and subsequent cotton production caused the destruction of the last 300,000 hectares of dry forest that remained in that country. In Central America generally, the most deforested and contaminated regions are those devoted to cotton farming. Unlike coffee, the farming of which substituted the high forest with a combination of coffee plants and larger shade trees, cotton created an ecologically unstable monoculture heavily dependent on agrochemicals and an escalating conflict between farmers and insects.

When, for political reasons, the North American market was closed to Cuban sugar in the sixties, Central America expanded sugar production by developing more sugar cane plantations and constructing larger and more modern sugar mills. This activity also concentrated in the Pacific coastal region of Guatemala, El Salvador, Nicaragua, and Panama.

Since colonial times, extensive cattle ranching has been an important activity, its epicenter being the drier plains of the Pacific, where natural grasses were abundant: Choluteca in Honduras, Granada and

León in Nicaragua, Guanacaste in Costa Rica, and Chiriquí and Los Santos in Panama. In the twentieth century, the growth of urban demand within the countries and the opening of new external markets boosted cattle breeding, particularly after the 1950s. As happened with coffee growing a hundred years earlier, the expansion of cattle ranching became a governmental priority. The lending policies of the national and international banks greatly contributed to the contemporary expansion of the pasturelands at the expense of the tropical forests. From the plains of the Pacific and the mountainous interior, cattle ranching began to make inroads into the wetter Caribbean.

By the end of the 1980s, 65 percent of the land under agricultural use in Central America was covered by pasture; of the rest, 16 and 19 percent, respectively, were occupied by export crops and basic grains. At the same time, between 1950 and 1980, the number of cattle multiplied from 4.2 to 9.5 million.

The Civil Wars (1973–90)

The three decades following World War II were a period of economic boom and social transformation. National economies grew at an annual rate of 4 percent, a growth rate higher than that in other parts of Latin America. An industrialization process was begun, and considerable resources were invested in infrastructure (fig. 7–3) and agricultural reform. All studies indicate a progressive improvement in income and living conditions for most national populations. Societies also became more complex and a middle class, almost absent previously, began to grow and prosper.

Then, abruptly, with the oil crisis of 1973, Central America was drawn into a dark period of economic depression, social dismemberment, and, in some countries, civil wars that would last until 1990. During these terrible years, and above all in the 1980s, often called the Lost Decade, large sections of the population experienced impoverishment. Production in many countries decreased to a level similar to that of the 1960s, and most people's incomes shrank to less than what they had been 20 years before. The production and availability of food decreased, and unemployment reached unprecedented levels.

These catastrophic changes were exacerbated by internal structural problems. Economic growth was accompanied by an increased concentration of landownership into fewer hands. In 1976, there were 1.25 million farms in Central America covering a total of 18 million hectares. Farms of fewer than ten hectares represented 80 percent of the total

7–3. An example of the investment in state infrastructure during the growth period of Central America from 1945 to 1975: El Cajón Dam, Honduras. Photograph by Stanley Heckadon-Moreno.

number of farms in the region but accounted for only 10 percent of the farmland. On the other hand, farms of more than 200 hectares, representing only 8 percent of the total number of farms owned, covered 70 percent of the land under use. There was also an increase in the number of landless families and of those who owned plots so tiny that they were not able to produce enough food to sustain themselves. Land concentration is particularly acute in the Pacific coastal region, where most land is held by either agro-exporting companies or large cattle ranchers. In other areas, for example, the Indian lands of the western high plateau of Guatemala, land is constantly split by inheritance into increasingly smaller fractions. Generally, the poorer peasants and Indians tend to occupy the more marginal and environmentally fragile lands.

During this period, the agro-exporting sector underwent major technological transformations, but subsistence farmers were left out of such support services as agricultural credit and technical aid. Most continued using traditional tools, like the axe and the machete as well as fire. These subsistence peasants, unable to be absorbed by the urban manufacturing sector or by the modern agricultural plantations, emigrated toward forest zones along colonization frontiers.

Another external factor in the crisis was the lack of justice in the relations between the industrialized countries and agricultural producers like the nations of Central America. Each year agricultural products are worth less and subject to tariffs, while the prices of imported manufactured goods increase.

All Central American countries are heavily indebted to international lending institutions and foreign governments. The external debt surpasses 40 billion dollars. The yearly interest payments reach awesome proportions. The enormous financial resources that governments have to set aside for meeting these payments reduces their capacity to provide the elementary social services necessary for their growing populations.

During the Lost Decade, there was also open warfare in Guatemala, El Salvador, and Nicaragua. Militarized regimes developed in Panama and Honduras. The United States and the former Soviet Union spent hundreds of millions of dollars and rubles in supplying the weapons and the training required by the armies of their respective allies. Within the countries, defense expenditures absorbed the lion's share of the national budgets. At the height of the conflict, there were 1.5 million armed men in five of the seven Central American countries, including armies, paramilitary groups, guerrillas, and their support groups. The casualties of the different civil wars exceeded 250,000, the wounded more than 400,000. On the missing there are no reliable numbers. Guatemala alone is estimated to have 50,000 missing persons. Some 3 million Central Americans became refugees, displaced internally and outside the region (fig. 7–4). It was a diaspora without precedent in Central America. Violence led to a massive flight of national capital and brains.

Violence overwhelmed national institutions and broke ethical values. For example, in the Usumacinta region of the Petén, Guatemala, the war took a dreadful toll between 1980 and 1987. Whole villages were trapped in the confrontation between the army and the guerrillas. One portion of the population became displaced persons inside Petén or other parts of Guatemala, and thousands exiled themselves into Mexico. Torture became a common practice, and the lovely Usumacinta River a dumping place for cadavers.

Presently, the civil wars of Guatemala, El Salvador, and Nicaragua have ended. Nonetheless, consolidating peace will be as demanding a task as war itself. In northern Nicaragua, for example, various armed groups have sprouted up, operating without control. These bands fall into four categories: radical leftist groups wearing Libyan gray and white uniforms who reject the peace treaties and assassinate right-wing politicians; former Sandinistas, wearing olive green uniforms with red and black handkerchiefs; well-armed former Contras who use North American weapons; and common bands of thieves neither well armed

7–4. A refugee camp in El Salvador after the civil war of the Lost Decade, 1980–90. Photograph by Stanley Heckadon-Moreno, with permission of ACNUR, United Nations.

nor well uniformed. In northern Nicaragua, the government periodically has to mobilize thousands of regular troops to force the armed bands to surrender their weapons.

But the greatest cost to be paid by most of the Central American countries' societies is not the huge material damage of the war. It is the social impact. The violence has had a profound effect on the individual and collective psyche, and the wounds are deep. They will take a long time to heal. Violence, particularly civil war, sustained through long years has created profound social traumas and a general suspicion and distrust of strangers.

The most positive change since the war and political violence have stopped has been the greater disposition of peasants to participate in their own self-help organizations and to seek to control their own destiny, at least locally. In urban areas the conflict generated a widespread crisis of the traditional political parties and their leaderships. This in turn has led to the emergence of new organizational structures and types of leaders.

Structural Adjustments of the 1990s

The 1990s have seen the end of one era and the beginning of another. During the past 20 years the prevailing model for development in most of the Central American countries was "military populism," in which the state machinery and military interference in national life grew enormously. Development was implemented from the top, the state assuming the role of producer. Although it achieved a few successful goals, it resulted in a gigantic external debt for most countries and huge bureaucracies.

It is difficult to predict what the new model for development will be like. In the current debate on the course to be taken by Central American countries, several issues hold crucial implications for the future. One involves the restructuring or drastic reduction of the state bureaucracy, selling of state-owned enterprises, and the redefinition of the manner in which the state will operate. In the future, the state may act as the promoter of private initiatives. In this connection, the integration of the Central American countries into a new economic federation takes on renewed meaning.

For the agricultural sector and the poor farmer above all, the situation is loaded with dangers and opportunities. In the coming years, the institutional framework will undergo changes, and institutions created to promote rural development will disappear or change drastically.

A positive factor in most Central American countries is the gradual spread of peace and economic recovery, which has translated into a new sense of optimism, undreamt of a few years ago. But peace and economic reconstruction increase pressure on the natural resources.

The Modern Colonization Frontiers

The cosmology of the ancient Indians held that order arose out of the harmony among gods, nature, and man. But the mestizo culture of Central America, strongly influenced by Judeo-Christian principles, operates on the premise that nature exists for humans to conquer.

Governmental development strategies of the 1960s and 1970s urged the need to incorporate the jungle into the national economy. Tropical forests were considered an obstacle to economic development and a symbol of national backwardness. Deforestation is the most radical transformation of nature brought about by existing patterns of economic development (fig. 7–5). In 1950, three-quarters of Central America was still covered by forest. Today, only 30 percent remains forested. Deforestation has devastated all types of forests, dry and wet, upland and lowland. In the past five decades man has destroyed more forest than in the previous 500 years. If present trends continue, by early in the next century the only surviving forest will be inside protected areas or the Indian territories. Presently, deforestation is estimated at almost 376,000 hectares per year.

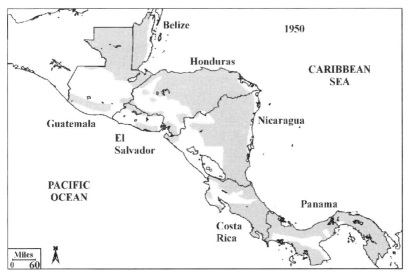

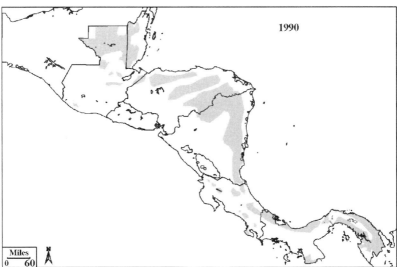

7–5. (*Top*): Extent of deforestation in Central America in 1950; (*bottom*): deforestation in 1990. Gray areas show presence of forests.

The driving forces in this drama of sweeping land use changes are poverty and technological backwardness, and the main actors are thousands of humble farming families, both campesinos and Indians, from the Pacific Coast and central mountainous zones. These are the most populated and ecologically degraded areas, and they suffer from the severest problems of landownership in Central America.

Traditionally, the zones that are subject to colonization on the Atlantic side have been inhabited by ethnic minority groups, Indians, and Blacks, with cultures well adapted to the forest and to the humid environment along the shores of the great Caribbean rivers and coastal lagoons.

In addition, these forests have been somewhat protected by their poor soils. The Caribbean slope contains 80 percent of the total protected surface area of Central America and 90 percent of the total timber-producing areas for the internal markets of Guatemala, Honduras, Nicaragua, Costa Rica, and Panama. Deforestation results in a massive extinction of the complex tropical fauna and flora. It also translates into deep and disturbing economic and social changes.

Once colonized, the areas become covered by pasturelands for cattle ranching. The cycle starts with poor soil on which the productivity of the cultivation of basic grains drops rapidly after clearing and burning. Once yields drop, after two or three years, farmers plant grass, which is burned every year in the dry season. The effects on the soil are deleterious, and within a few years the productivity of cattle ranching also decreases. Thus, the existing farming system is neither economically nor ecologically sustainable, and it does not provide most farmers with enough to live on. When they are no longer able to support their families, they sell their properties to bigger cattle breeders, so land soon concentrates in a few hands. In other cases, the area is simply abandoned. If the soil is not too degraded, secondary forest slowly develops. In Costa Rica, for example, there are more than 400,000 hectares of such growth. Even though the colonization frontier first produces small and medium-sized cattle breeders, land is ultimately concentrated into the hands of only a few owners. Cattle-breeding zones generate few jobs and tend to force the population to move elsewhere.

In the Pacific and central mountainous zones, pressure on surviving forests is mostly due to the use of firewood as the main source of energy. Most poor people in towns and cities, unable to pay the high cost of electricity and gas, cook with firewood. On average, firewood in Central America supplies half the total energy consumed: a rate of 1 cubic meter per person per year (fig. 7–6).

A second major environmental impact of economic development, one that remains largely invisible and goes unrecorded in national accounts, is the millions of tons of soil washed from the central mountain ranges during the rainy season and discharged by creeks and rivers into the sea (pls. 15, 16A). This silent, daily process goes on year after year, each generation adding to the problem, the present one contributing far more to this deterioration.

7–6. A major cause of deforestation in the Central American highlands and the Pacific coastal zone is the cutting and collecting of firewood, which provides half of all the energy for domestic consumption and cottage industries. Here a stockpile near León, Nicaragua. Photograph by Stanley Heckadon-Moreno.

In coming years, water will become a major ecologic problem in Central America. Apart from the natural droughts of the dry season, deforestation and degradation of watersheds that supply water to population centers are resulting in water shortages (fig. 7–7). This problem is most acute on the Pacific side. With vision and great sacrifices, some countries have engaged in major electrification programs. However, the useful life of many of these projects is threatened by sedimentation: Chixoy in Guatemala, El Cajón in Honduras, and Bayano in Panama are cases in point. For example, in Nicaragua on a rainy day, 36,454 tons of sediment passes down the Malacatoya River on its way to the July 29 dam near Lake Managua.

The culmination of these adverse economic and ecological impacts is to create an extraordinary social process, namely, major *colonization frontiers,* as the displaced poor move inexorably to colonize and cut down the humid forest of the Atlantic coastal zone. From Petén in Guatemala to Darién in Panama, the agricultural frontier is expanding vigorously and destructively along a front that extends almost 1500 kilometers. This frontier destroys existing natural resources and fails to replace them with a viable livelihood for the immigrants.

The most aggressive forest colonization frontiers in Central America are the Petén in Guatemala, the Honduran Mosquitia, the Coco and

7–7. Scene at a government well, drilled to offset surface-water shortages in the dry season, on the plains of León, Nicaragua. These campesinos had traveled 11 kilometers to fill their oil drums for domestic water supplies. Photograph by Stanley Heckadon-Moreno.

San Juan River basins of Nicaragua, and the provinces of Bocas del Toro and Darién in Panama.

Petén, Guatemala

Consisting of 36,000 square kilometers, Petén is the largest department of Guatemala, almost as large as Belize and El Salvador combined. Its soils, like those that typically characterize an extensive karst limestone terrain, are poor, clayey, and prone to drainage problems. This terrain is interrupted only by the Maya Mountains. It is Guatemala's most dynamic colonization front. Three-quarters of the new farmland in this country is being opened in Petén.

Searching for land, 300 new immigrants reach Petén daily from other parts of Guatemala. Ten years ago, Petén had 40,000 people: today, the population exceeds 350,000, and native-born *Peteneros* are a minority. In the midst of the poverty prevailing in the region and the declining food production, it is salutary to recall that in this fragile environment the Mayans supported millions of inhabitants. Traditional Peteneros have always been a forest society with a culture, unique in Guatemala, based upon gathering of rubber and chicle (obtained from a resin-bearing tree and used for making chewing gum), harvesting of precious timbers, and extracting of "xate" and "fat" pepper. From 1890 to 1970, the commercial extraction of chicle for export dominated the economy, and the Petén is covered with old airstrips where chicle growers established themselves and where many of the present towns have grown up. During these times, cattle breeding, although dating back in Petén to the eighteenth century, was limited to the natural savannas and advanced very little on the forests.

Petén has a unique institutional history among the ongoing colonization fronts of Central America. From 1958 to 1990, a state development corporation run by the army and always headed by a general was responsible for the promotion of colonization and development. The central government wanted to settle landless people from the central highlands and the Pacific coast in Petén in order to achieve two goals: defuse social pressure on land resources in the more densely inhabited areas and establish a stronger Guatemalan presence in the thinly populated border areas with Mexico. This was intended to counteract what was perceived as the Mexican threat and the Belizean problem. Guatemalans remember bitterly Mexico's annexation of Chiapas, just as Mexicans carry memories of the loss of California and other territories to the United States.

The government development corporation parceled land in large holdings, mostly in southern Petén, and tended to prevent colonization in the north. The gradual consolidation of peace since the civil war, however, has now accelerated forest colonization; refugees that had gone to Mexico are returning, and farming is expanding. Much timber is now being extracted, legally and illegally, the most valuable of which is smuggled into Mexico and Belize, where prices are higher.

Immigrants have now extensively deforested the southern Petén and degraded the soil. Some estimates indicate that 40 percent of the soils are now heavily deteriorated. As a result, migrants are now moving northward toward the Mexican border and eastward into the pine forests of Chiquibul in the Maya Mountains. As these internal migrants seek new land, their situation has been aggravated by the appearance of *hilo,* a serious fungal infection affecting the bean crop. The disease has become so rife that the Petén is no longer able to support the agricultural needs of this growing population.

Northern Petén holds half the forests of Guatemala, including the great Maya Biosphere Reserve comprised of more than 1.5 million hectares, and the most important and numerous array of Maya sites in Central America, including Tikal. They constitute a peerless ecological and cultural patrimony.

The Maya Biosphere Reserve is a typical case of the growing confrontation in the settlement frontiers of Central America between local rural communities and the move to conserve biodiversity through protected areas. These reserves, copied from the experience of the developed countries, are being used in the management of natural resources. When the Maya Biosphere Reserve was decreed in 1990, its boundaries did not follow the contours of the terrain but were laid out in straight lines, and many communities lay within the new protected area. Conservation authorities were unclear as to what policy would be followed regarding these settlements. Park rangers were told that the Maya Biosphere was an area for total preservation, untouchable, to be kept like a museum. The first practical measures taken by the National Council for Protected Areas to manage the area were to install control posts and to make breaches marking the boundaries. These actions, taken without explanation to the local communities, threatened the security of their lands, houses, and crops. Not surprisingly, three government control posts were burned down by irate villagers.

Presently, three colonization fronts are moving on the northern forests and archaeological sites of the Maya Biosphere Reserve, all triggered recently by the construction of major roads. The immigrants who colonize Petén along these roads start as corn planters, but their desire is to become cattlemen, a cultural pattern deeply rooted among the traditional cattle-ranching mestizo peoples of Guatemala.

The immigrants make the forest retreat by planting pastures. However, in summer there is very little forage because of the poor quality of the soils and the practice of burning the pastures. After three or four years, with soils deteriorating, the ranchers need to clear more forest for new pasturelands. As in all Central American colonization fronts, small farmers are displaced by small and medium-sized cattle ranchers who eventually are themselves displaced by even larger cattle owners who take over most of the land.

Petén is a very large region with rudimentary communications and transportation systems. There are problems with guerrillas and locally organized bandits, some smuggling fine lumber to Mexico and Belize, others running drugs or illegal immigrants toward the United States. There has been also the overwhelming presence of the Guatemalan army.

In spite of these difficulties, Petén has become a magnet for national and international environmental and developmental organizations seeking to halt the expansion of the agricultural frontiers and to save the surviving forest and the Maya historical sites. About thirty such organizations have projects in Petén, only one of which is headed by a local person. Intense, sometimes bitter competition exists among these organizations; there is hardly any coordination of activities. Little of the foreign money for sustainable development projects seems to reach the local communities. Rather, it appears to support a growing nongovernment organization bureaucracy and their rental of houses, offices, four-wheel-drive vehicles, and computers. Elaborate workshops, seminars, and brochures are produced, often aimed at potential external donors. Some projects also buy community participation in environmental projects by giving away food. Despite these efforts and the substantial amount of money invested, progress appears to have been meager.

Perhaps one of the most hopeful projects is taking place at Bethel, one of fourteen cooperative societies established by the government of Guatemala in the 1960s, at the border with Mexico. Many poor southern campesinos answered the government's promise to give them forest parcels, houses, and seeds at Bethel. The first settlers arrived in 1968, but the promised help never arrived, and the people, like the Maya settlers thousands of years ago, survived the first months by eating an edible fruit from a small tree called *ramón*.

The cooperative has about 5500 hectares of land, of which 2740 hectares are common forests, mostly within the nucleus of the Maya Biosphere Reserve. With help from national and international organizations, there is ongoing work on a pilot plan to manage the natural forests, based on the model developed for the common forested land of Quintana Roo, Mexico. The objective is to show the peasants that forests are a resource which, if well managed, can be an important source of income and new jobs. If successful, Bethel will serve as a model for other cooperatives in Petén that still have forests.

Several organizations in the Petén are trying to promote the use of bean fertilizer for soil enrichment and weed control. The objective is to reduce the use of chemical inputs, develop more resistant genetic materials, and to cut by 75 percent the amount of soil annually required by a peasant family to grow maize.

La Mosquitia, Honduras

Honduras is the second largest country in Central America, with 112,000 square kilometers. Three-quarters of its territory is mountainous, and most of its soil is of low fertility. The best soils are found in the narrow valleys of the interior rivers, in the coastal plains of the dry Pacific, and along the humid Atlantic coast, where the banana plantations are concentrated.

The most important recent change in land use has been the expansion of pastures, which rose from 822,000 hectares in 1952 to 2.6 million hectares in 1990. Pasturelands represent 50 percent of the total area dedicated to agricultural use. During this period, the area dedicated to basic grains has been reduced. Presently, forest covers 50 percent of the country. Deforestation is estimated to be more than 100,000 hectares a year. By the year 2000, the surviving forests will be restricted to the protected areas and Indian territories.

Southern Honduras has many similarities with other zones of population emigration on the Pacific coast of Nicaragua and Guatemala. Poverty, land concentration, and environmental degradation are fueling a rural exodus toward the towns or the distant colonization fronts. Previous state efforts in the 1960s and 1970s in conservation and natural resource management have largely failed owing to a top-down approach and the failure to involve rural communities in planning or implementation. In the departments of Choluteca and Valle, hundreds of kilometers of painstakingly built stone fences to control soil erosion now lie abandoned and overrun with cattle.

Mosquitia occupies more than 16,000 square kilometers and has some 50,000 inhabitants belonging to five groups: Miskitos, Garífunas, Tawahkas, Pechs, and mestizos. At present there is a massive immigration of mestizo colonists from the southern departments of Choluteca and Nacaome and also from Olancho. Colonists move into territories occupied by Indian groups and destroy the forests.

Today, Dulce Nombre de Culmi in Olancho serves as a reception center for immigrants looking for forested land. From here they move into Indian-occupied territories south of the great Río Plátano Biosphere Reserve (535,000 hectares) and the reservation of the Tawahka-Sumo Indians (233,000 hectares), destroying forest as they go (see pl. 16B).

The massive penetration of immigrants from the drier departments of southern Honduras was facilitated by the construction of a major road from Catacamas to Culmi and then into the basin of the Wampú River, where the buffer zone for the Río Plátano Biosphere Reserve begins. A few years ago this hilly terrain was covered by extensive forests of Honduran pine. Today, the landscape is mostly pasture with surviving pockets of degraded pine groves. The construction of the road to Culmi was a national priority of the government of former president Jose Azcona.

More than in most countries of Central America, forestry plays an important economic role in Honduras. In the new colonization frontier of La Mosquitia, the opening of the Culmi road gave rise to a brief economic phase of timber extraction, which skimmed the most valuable wood from the forest. Lumber dealers with heavy equipment opened a spiderweb of roads to transport the logs.

Throughout Central America a large proportion of the legislators in the national assemblies hail from rural constituencies and are cattle or lumber men, historically favorable to the expansion of the frontier. Cattle lobbies are among the most powerful political pressure groups within the country. When confrontations have occurred between cattle ranchers or lumbermen and forest-dwelling Indian and Black communities, the legislative powers usually have sided with the colonists.

As in all colonization fronts the most characteristic sound along the Culmi road is not the song of birds but the whine of chainsaws. This instrument has revolutionized forestry, increasing many times over the capacity to extract timber. In Honduras and other countries, owning a chainsaw is now an aspiration of most campesinos.

Culmi and the other hamlets along the new road resemble lawless frontier towns, with a weak representation of national institutions and few local organizations. The installations of the Natural Resources Secretariat in Culmi have been closed for lack of funds, and there are no agricultural extension or agricultural credit services in the district.

Following the phase of lumbering comes the cycle of cattle raising with the usual land concentration into the hands of fewer cattle ranchers. This structural problem persists regardless of the agrarian reform projects made by the state. Poverty and environmental degradation then force the rural population toward the towns or to the humid forests of the Atlantic side. The road to Culmi runs north and parallel to the lovely Sierra de Agalta Mountains. A few years ago the Sierra de Agalta National Park was established. It is a small protected area without enormous biodiversity, but it is of critical local significance. In the 65,000 hectares of surviving forests of this mountain range arise six rivers that are the main supply of water for the densely populated valleys of Catacamas and Juticalpa in Olancho, a department from which many colonists move to La Mosquitia. From the valleys of Olancho, cattle ranching and maize fields move into the heart of the park. If this advance is not contained, the park will disappear within ten years. Inside the Sierra de Agalta National Park there are 132 villages and 1200 small coffee-growing families. These farmers engage in clearing and burning, practices that cause serious erosion problems. To deal with the problems of conservation and development in this critical area there are only four park rangers and one technical assistant.

In Honduras, the Department for Protected Wildlife Areas oversees 104 protected areas, most of them created within the past decade, with thirty-two ill-equipped forest rangers. The Honduran government, like all others in the region, faces dire fiscal limitations in providing the means with which to execute its growing mandate efficiently. The scarce resources are consumed by salaries, leaving little for equipment, infrastructure, and maintenance. In Honduras, as elsewhere in Central America, the field offices of the national parks are equipped with unreliable vehicles, outboard motors, and radio systems and with outdated maps and satellite images. Policies and projects of the environmental agencies are frequently obstructed or abandoned through political interference and the resulting changes in staff.

The Coco and San Juan River Basins, Nicaragua

In 1950, Nicaragua had some 8 million hectares of forests. Today, it has only half that amount. The famous pine groves of the Nicaraguan Mosquitia have been reduced by annual fires and the exploitation of the best trees to half a million very degraded hectares.

The dry forests of the Pacific of Nicaragua are almost extinct, in great measure because of the expansion of the cotton fields since the 1950s. Initially, cotton farming displaced the peasants and their basic grain production from the Pacific and central mountainous zones into the rugged terrains further east. Later, they would be displaced further, most of all by the boom in cattle ranching. In 1950 Nicaragua had 800,000 head of cattle; by 1980 it had reached almost 4 million, and cattle ranching had penetrated to the Atlantic coast.

In most parts of the older colonization frontiers of the interior highlands of Nicaragua, 85 percent of the soil is no longer adequate even for cattle breeding, and much of it lies seriously eroded between large gullies and rocky outcrops. The two important modern colonization frontiers are both on the Atlantic coast. The first is the basin of the Coco River on the border with Honduras, enclosing the great natural reserve of BOSAWAS (a word derived from the first letters of the rivers Bocay, Sang Sang, and Waspuk), which comprises 8000 square kilometers of broken terrain and tropical forests. The other frontier is in the San Juan River basin on the Costa Rican border and contains the Indio-Maíz Biological Reserve, with 295,000 hectares.

BOSAWAS is the largest forested zone of Nicaragua but also the most dynamic colonization front, and it is the area with the least institutional presence. Not surprisingly, it was also the region in which the most violence took place during the civil war. For Nicaragua the stabilization of this agricultural frontier is crucial. It is the region with the greatest biodiversity, and its forests cover a very broken mountainous knot where the six largest rivers of the country originate. These rivers could potentially supply 80 percent of the national hydroelectric power. Furthermore, these rivers flow into the great lagoons, mangrove swamps, cays, and reefs of the Atlantic coast, where they contain the country's most important marine and coastal resources. Uncontrolled colonization of the steep lands at the headwaters of these rivers will have a devastating effect on all of these vital maritime and coastal systems.

BOSAWAS has some good agricultural soil, especially in the alluvial river valleys, and these lands are very attractive to settlers. The area is ethnically and culturally complex, Sumo and Rama Indians as well as Miskitos residing there along the Coco and Bocay rivers. During the past 3 years, a growing number of Spanish-speaking mestizo colonists have arrived from the drier departments of the Pacific and the central mountainous regions, although there is no recent accurate census because of the civil war.

The first mestizo peasants arrived in the 1950s, after the development of cotton growing had expelled them from the Pacific coast. A lumber company then opened a road that provided the only link with the rest of the country. In the 1970s, the region became a stronghold of the Sandinistas during the protracted struggle against the dictatorship of President Anastasio Somoza, and again in the 1980s during the civil war it was the most important operational base on the Honduran frontier. Most of its population was relocated by the Sandinista army away from the border inside Nicaragua or took refuge in nearby Honduras. Since the conclusion of the conflict the refugees are returning. Among them are Sumo Indians who had taken refuge in Honduras or had joined the Contras, mestizo ex-servicemen of the national resistance, and landless peasants, who came to know the area well during the war.

In southern Nicaragua lies the great and historic San Juan River, the largest hydrographic basin of Central America, which occupies some 40,000 square kilometers and has an annual rainfall of more than 5500 millimeters. On the northern shore of the San Juan lies an extensive group of protected forested areas collectively known as SIAPAZ (the acronym means yes-to-peace): the Río Indio–Río Maíz Biological Reserve (300,000 hectares), the Wildlife Refuge of the San Juan River Delta (100,000 hectares), and the Guatusos Wildlife Refuge (100,000 hectares).

Unlike BOSAWAS, the SIAPAZ project is not home to any Indian groups. The colonists are mestizos from older colonization movements and families that fled to Costa Rica during the war. Ever since the war ended there has been a great movement of refugees looking for land.

Bocas del Toro and the Darién, Panama

The main areas of agricultural colonization in Panama are located at opposite ends of the country: the Province of Bocas del Toro in the west and the Darién in the east. Like Petén, Mosquitia, and the Atlantic zone of Nicaragua, Bocas del Toro has long been isolated from the Pacific coastal zone, but this has rapidly changed since the construction of a highway parallel to the transisthmian pipeline that transports crude oil from the Pacific to the Caribbean coast. Taking advantage of the new road, a growing number of mestizo settlers and Guaymí Indians from the more densely populated Province of Chiriquí, on the Pacific side, have begun to settle in Bocas del Toro.

Bocas del Toro offers a complex ethnic and cultural panorama. Its traditional Indian groups are the Guaymí, who historically have inhabited the valley of the Cricamola River and the Valiente Peninsula; the Teribe, living chiefly along the Teribe River; and the Bribri, who are located on the Yorkin River. Black English-speaking Creoles live in the coastal lagoons and islands of the archipelago, having arrived early in the nineteenth century from the islands of San Andrés, Providence, and Jamaica. More recently, Spanish-speaking mestizos have become a growing element of the population.

Bocas del Toro resembles Petén in having an important number of Indian immigrants. Guaymíes have come from the provinces of Chiriquí and Veraguas on the Pacific side, where a rapidly expanding population faces agricultural and ecologic deterioration and a growing deficit of firewood and timber. Guaymí from the Pacific zone also practice extensive cattle rearing. In some areas, the Pacific-side Guaymí are invading territories historically belonging to the Teribe and Bribri Indians. Bocas del Toro now contains 37 percent of the primary forest of Panama, but there is unplanned and uncontrolled timber extraction by the Guaymí and Teribe Indians and by the Creole and mestizo campesinos.

The traditional Black and Indian communities of Bocas del Toro are also facing a crisis because their principal source of money, the cocoa bean, has seen a marked drop in price on the international markets and has been devastated by an infection of the fungus *monilia*. This decrease in their income due to loss of cocoa has driven the forest communities to turn to lumbering even in the protected areas.

With the advance of the colonists from the Pacific side, the Teribe, Bribri, and Guaymí from Bocas del Toro are determined to formalize the demarcation of their territories, a process that threatens to end in strong interethnic conflicts. In the frontiers of settlement in Honduras, Nicaragua, and Panama, parallel to the process of colonization, there is an ongoing movement of the forest-dwelling communities to obtain security of land tenure and official recognition of their internal political organizations. These aspirations are articulated more vehemently by the Indian than by the Black communities.

One of the most important protected areas in Panama and in all of Central America is the Parque Internacional de la Amistad, which contains 207,000 hectares. La Amistad is a binational park shared by Costa Rica and Panama. Owing to its extraordinary biological diversity it has been designated a biosphere reserve and world heritage site. Its main buffer zone is the Palo Seco protective forests, 250,000 hectares of very wet tropical forests. In Palo Seco lie the headwaters of the Teribe and Changuinola rivers, which hold 90 percent of the hydroelectric potential of Panama.

Minimizing the colonization of Palo Seco is vital for the survival of La Amistad park. The Teribe and Bribri Indians are the custodians of the two main entrances to the park: the Teribe and Yorkin Rivers. Other small but ecologically unique forests are found on the islands of the archipelago of Bocas del Toro. Because of their great potential for scientific and ecological tourism, these islands have been characterized as the Galápagos Islands of the 21st Century. Within them lies the Bastimentos Island Marine Park. For the Indian and Black communities of these islands the main source of income has been turtle and lobster fishing. Overfishing is exhausting both species. Only the successful development of nature tourism would seem to offer an alternative to the eventual destruction of the unique resources of the Bocas archipelago.

The Darién comprises 16,000 square kilometers. The population consists of three major ethnic groups: Indians, including the Kuna, Emberá, and Wounaan; the Black communities, known as *darienitas* because they have been born in the Darién; and the Spanish-speaking mestizo settlers from the interior of Panama, particularly the provinces in the dry Pacific coastal zones.

Whereas the mestizo campesinos employ horses as their main means of transportation, the Indians and Black darienitas, groups attuned to centuries of living in the forested habitats of the large rivers of Darién, use the *piragua,* or shallow dugout canoe, a cultural artifact that symbolizes their amphibian cultures.

The ethnic equilibrium of Darién shifted rapidly when the Pan-American Highway to Colombia was constructed in the 1980s, turning Darién into the principal Panamanian colonization frontier. In 1983, the highway reached Yaviza, a small town on the Chucunaque River, and stopped. The remaining 75 kilometers of forest that separate Yaviza from the Colombian border have never been paved; this stretch is called the Darién Gap.

Following the opening of the Pan-American Highway many campesino settlements sprung up in Darién. The most important is Metetí, which serves as the headquarters for several state agencies and the largest stores. Metetí functions as a springboard for colonists setting off for the remotest parts of Darién adjacent to the reservations of the Kuna and Emberá Indians and the Darién National Park. Along the highway there is extensive deforestation and the expansion of cattle breeding, as the forest has given way to degraded pasturelands. The cultivated areas are few, and the short-lived timber extraction boom is now collapsing as the valuable timber is exhausted.

The surviving pockets of forest tend to be located inside private farms. In Darién, as in other colonization fronts in Central America, many settlers with large farms have pockets of natural forests. Taken as a whole, these surviving forest plots add up to thousands of hectares. Forestry institutions in the frontiers, however, are unable to provide technical advice or supervision for the management of these forests. As in Petén, Mosquitia, and BOSAWAS, 90 percent of the working time of state forest technicians in the Darién is spent in inspecting, granting permits, and overseeing timber extraction, leaving little time for forestry research and management.

Faced with the aggressive advance of the mestizo colonists, the Emberá and Wounaan Indians have organized to defend their land. Such action has resulted in the forging of a strong political organization, the Emberá-Wounaan General Council, and the establishment of the legally recognized Emberá Indian *comarcas,* or reservations, which encompass 25 percent of the Darién. The relation of the autonomous Indian comarcas with the protected wildlife areas is complex because they extensively overlap national parks and protected areas. For example, it is estimated that 297,000 hectares of the Darién National Park overlap territories of the Emberá and Kuna Indians.

Indians and darienitas are also experiencing an agricultural crisis. Family monthly income in Darién is about $82, and studies by the Panamanian government show that 75 percent of the people of Darién live below the poverty line. The main reason for the crisis has been the destruction of the plantain farms, the main cash crop of the Indian and Black communities. Until recently, Darién was the largest producer of plantains in Panama. Emberá, Wounaan, and darienitas cultivated thousands of hectares on the fertile alluvial terraces of the main rivers. The plantains were sold to intermediaries, who took them by sea to Panama City. Shortly after the construction of the Pan-American Highway, in 1980, Black Sigatoka fungus appeared in Darién, and in a few years the disease reduced plantain production by 75 percent.

Another factor in the agricultural crisis can be traced to the economic measures imposed on Panama by the international financial organizations. These adjustments have led to the closing of the agricultural purchasing posts of the National Marketing Institute in the distant riverine villages. These posts offered local producers the only alternative market for their rice and corn production.

Such extreme poverty as that afflicting the native population of Darién has immediate environmental repercussions. In the town of Yaviza, for example, representatives of the Emberá communities are constantly applying to the Ministry of Natural Resources for extraction permits to sell standing trees at low prices to lumber dealers. For these communities the sale of timber is rapidly becoming the main source of cash.

In 1980, the Darién National Park was created. Measuring 5790 square kilometers, it is by far the largest protected wildlife area in Panama, covering 30 percent of the territory of Darién. To patrol this vast area there are seventeen rangers. Many Indian and Black communities were enclosed by the new protected area. Rules of the park decree that it is an area of biological protection and that hunting, fishing, felling of trees, clearing of land, and timber extraction are prohibited. As an experienced park ranger in Darién recently told me, "We are repressing the people, and that is going to explode."

It is possible that in the near future Panama and Colombia will complete the construction of the Pan-American Highway across the Darién Gap. The road will unite the Americas—but it will also certainly have a disastrous impact on the Darién National Park and on the cultures of the forest-dwelling Indian and Black communities.

Sociology of the Colonization Fronts

Colonization frontiers involve migrants from regions marked by serious social and environmental problems. Some come from areas where land is concentrated in a few hands, others from areas where the plots are too small to be viable. But in general they come from regions that have historically been dominated by extensive cattle ranching.

In their places of origin, the colonists have become outcasts, and they see the frontier as the promised land, a land where they can begin a new life. Because the main aspiration of most Central American campesino migrants is to become cattle ranchers, however, they have little affinity for the forest, considering it a hostile environment and, in a sense, their enemy. Progress and development mean the need to cut down the forest quickly. This attitude usually puts the settlers in opposition to the traditional forest-dwelling groups.

Behind the label *agricultural frontier* in Central America there is great physical and social diversity. Yet there are several common themes. First is the overwhelming poverty of the majority of the people: the average annual per capita income is about $100. Second is the hopeless, nonsustainable method of land use, which destroys the existing biodiversity and replaces it with a deteriorating landscape that cannot continue to support its people, so that they are forced to migrate in the search for more forest to clear. A key challenge of the future is to intensify scientific research and development so as to change the technology used by the rural producers. Slash-and-burn agriculture and extensive cattle ranching have to be transformed because they are currently unsustainable. They are also compounded by a rapidly growing population and a diminished base of natural resources. Governments will need to intervene with subsidies to educate farmers and implement improved techniques. The central and crucial challenge of contemporary Central American societies is to transform the present campesino farming systems so that they become sustainable within defined limits and cease to deteriorate progressively and inexorably the national resources.

Third, the Indian and Black forest-dwelling communities are the custodians of the majority of the remaining forest wealth of Central America. Pressed by their material needs and growing populations, however, these communities are increasingly mining their forests, selling the most valuable timber for short-term cash benefits.

The demand for forestry products, especially timber, will almost certainly continue to grow. It is therefore extremely important that forest-dwelling communities in the colonization fronts be trained in the management of their natural resources. They need to know how to establish sustainable forest management plans and to profit more from development and commercialization. Village forestry training must include the selection, treatment, and commercialization of seeds of native timber-yielding trees for reforestation of already degraded areas as well as research and development of nontimber forest products.

Fourth, the declaration and establishment of protected natural areas in the distant frontiers of settlement, considered an accomplishment by the scientific, national, and international environmental groups, are seen with different eyes by the poor rural communities. Increasingly there are confrontations between park rangers, campesinos, and Indian communities, who feel that the land use restrictions enforced by the parks worsen their poverty. The enforcement of existing land use regulations seems to many to be a policy of sticks without carrots. Today, a basic dilemma in Central America is how to reconcile the laws of the protected areas with the subsistence needs of the poor who inhabit them.

Generally, protected areas have been established without prior evaluation of the local socioeconomic problems and without provisions for the necessary means to protect them effectively. Landownership regulations in the parks are not clearly defined. Most protected areas lack management plans, and if they do exist, seldom have they been discussed with local communities.

Fifth, throughout Central America an extensive range of evangelical sects is rapidly expanding its influence. These religious movements generally adopt a position of confrontation with the Catholic Church, the traditional religion of Central America. These sects are also mistrustful of traditional Protestant churches and of each other. Such religious fragmentation hinders the establishment and achievement of community goals. Some of the radical sects are now using their often considerable financial resources and growing religious influence to obtain political power. Most of these movements have shown little or no awareness of the demographic, agricultural, and ecological crises that the colonization fronts are rapidly provoking.

There are, however, some encouraging signs. Since the 1970s, and with greater intensity in the 1980s, there has been a rising level of ecological consciousness throughout Central America. People are more aware of environmental problems and their implications for the future. They are demanding greater accountability of governments and private enterprises. Groups seeking to protect nature and establish a style of more sustainable development are constantly increasing. In turn, public opinion is forcing politicians to change older national policies based on the myth that natural resources are limitless.

Slowly, countries have built their institutional capacity to deal with environmental issues, and in the national assemblies environmental commissions have been set up to enact a new legal framework for natural resources (see chapter 9).

These accomplishments include the establishment of the national systems of protected natural areas. In the 1970s there were fewer than thirty protected areas in Central America; today, there are more than three hundred, comprising more than 9 million hectares, 90 percent of which is concentrated on the Caribbean side of the isthmus. Initially, most protected areas were established by the central governments, but today local governments and the expanding environmental movement are major forces behind the creation of new protected areas.

Furthermore, in the deforested and degraded areas of the Pacific side and the central mountainous zone of Central America, a silent revolution is taking place: the cultivation of trees for commercial purposes. Until 1990 only 30,000 hectares had been reforested in all of Central America. Governments have again been the pioneers of reforestation, through the small forestry departments within the ministries of agriculture. Much of what was first planted has been lost through lack of proper management and local community participation. Indeed, in many instances, rural communities opposed early reforestation schemes imposed on them by national organizations without consultation.

Now, as a result of gradual changes in policies and incentives and growing market demand, an increasing number of private investors are planting trees on a commercial scale. Ironically, now that there is an awakening interest in reforestation, there is a scarcity of seeds from native tree species as deforestation destroys the source of the best timber-yielding species.

Another recent transformation in Central America is the process of decentralization, which has involved the gradual transfer of power and resources, including the management of natural resources, from the distant central government to the local municipalities. The strengthening of local government seems to be encouraging more local communities to participate in land use planning. Among the most important tasks to be tackled by local governments in the future is the protection, management, and reforestation of critical watersheds that supply water to the growing populations.

Obviously, the task of protecting and managing natural resources in Central America unfolds within the context of awesome socioeconomic challenges, including a dramatic demographic revolution and rising poverty. Thus, the development model to which Central American societies should aspire must seek social and economic justice. It is a race against time. In Central America a good share of the roots of violence lies in poverty. The new model must seek harmony between man and nature through the substantial conservation and appropriate exploitation of biodiversity.

CHAPTER EIGHT

Central American Indian Peoples and Lands Today

PETER H. HERLIHY

The indigenous peoples or Indians of Central America today are grouped into a series of "peoples" identified by their languages, traditions, settlements, and subsistence activities. Because the Indians' lands and cultures are threatened, their future may well depend on conservation initiatives to protect regional wildlands and endangered habitats that now more or less coincide with their historic homelands. Once Indians and forests covered the Central American isthmus. Today, both have restricted ranges that are increasingly threatened by outsiders.

Scholars debate the size of Central America's aboriginal population at the time of European contact. Spanish chroniclers observed significant populations, especially along the Pacific slope, but their precise numbers will probably never be known. Recent reviews of the documentary and scholarly evidence for the numbers of aboriginal inhabitants of the Americas in 1492 estimate a population of 5.6 million, distributed approximately as follows: 1 million in Panama, 1 million in Nicaragua, 500,000 in El Salvador, 750,000 in Honduras and Belize combined, 400,000 in Costa Rica, and 2 million in Guatemala.

Indigenous Populations Today

About forty-five distinct indigenous populations survive in Central America today. (The names of these Indian peoples vary; if more than one name is frequently used, I give the alternatives in parentheses after the first use of the name.) Estimating the size of these groups is a difficult task. Indian populations inhabit regions that are difficult to reach, and census-takers often greatly underestimate their numbers. Also, the definition of *Indian* is often quite subjectively based on the outsider's interpretations of language or material cultural features. According to a recent National Geographic Society/Cultural Survival survey, roughly 3.9 to 5.6 million indigenous people live in Central America today, accounting for 13 to 19 percent of the total population in 1990 of 29 million (table 8−1).

The strength of the native element differs from state to state. Guatemala is recognized for its Indian heritage, whereas Costa Rica appears European. However, although colorful Maya costumes attract much attention, other countries of Central America have considerable areas occupied by native peoples (fig. 8−1).

Belize

Archaeologists credit Belize with one of the earliest Mesoamerican village sites; the Maya were widespread by Classic times (A.D. 600−900) and had large populations in this region. They remained isolated from much early Spanish colonial development but were later reduced or totally removed, probably by exotic diseases. The present Maya settled here in the nineteenth century. Two groups, the Mopán and Kekchí, crossed into southwestern Belize from the Petén in the 1880s. The Mopán came in search of political freedom and agricultural lands, while Kekchí initially sought work on plantations. By the 1930s, the government had set aside 63,000 acres for Indian reserves. About 4400 Kekchí and 4000 Mopán live in the country today.

The mid-nineteenth-century Caste War between upper-class Spanish-mestizo and native Maya of northern Yucatán brought Maya south into Belize. The Yucatec once numbered 10,000 here, but in 1990 only some 6000 lived in about forty villages. Most Yucatec are highly acculturated, work for wage labor, and have lost the use of their language.

The Garífuna are a highly distinctive people who live along the coast from Dangriga southward. Not native to Central America, their populations derive from Black slaves who mixed with Carib Indians on Saint Vincent Island in the Lesser Antilles. They were exiled by the British to the Honduran Bay Island of Roatán in 1797. From there, they crossed to the mainland at Trujillo and spread along the coast over the next four decades, settling Dangriga by 1832. Although they are physically Negroid, their Amerindian lifeways, biological makeup, and use of indigenous language cause experts to consider them an indigenous

Table 8–1. Central American Indigenous Population, 1940 to 1990

Country	Population Total		Indigenous Population			
	1940[1]	1990[2]	1940[1]	(% Total)	1990[3]	(% Total)
Belize	55,000	190,000	13,134	(23.9)	25,200	(13.3)
Guatemala	2,380,000	9,200,000	1,309,000	(55)	2,963,400	(32.2)
El Salvador	1,744,535	5,250,000	348,907	(20)	500,000	(9.5)
Honduras	1,107,859	5,140,000	100,000	(9)	157,700	(3)
Nicaragua	900,000	3,870,000	39,400	(4.4)	94,800	(2.4)
Costa Rica	656,129	3,020,000	3,500	(0.5)	23,368	(0.8)
Panamá	674,403	2,420,000	55,987	(8.3)	196,700	(8.1)
Total	7,517,926	29,090,000	1,869,928	(25)	3,961,168	(13.6)

1 Source: Angel Rosenblat, 1954. *La población indígena y el miestizaje en América.* Tomo I.
2 Source: *World Resources 1992–93,* P. 246.
3 Source: *Research and Exploration,* Map Supplement, 1992.

group. They are coastal dwellers who live in clustered settlements that reach Plaplaya in eastern Honduras, with two settlements in Nicaragua. Over 11,000 Garífuna live in six villages in Belize, making them the country's largest indigenous group. Although they are culturally strong, colonists threaten their territorial security.

Indigenous populations in Belize struggle to maintain their lands and cultural identity. The National Garífuna Council and the Toledo Maya Cultural Council, representing Kekchí and Mopán, have been formed to help sustain their cultural integrity.

Guatemala

Guatemala remains the most Indian country in Central America today, having more than 3 million Maya speaking twenty different languages. The only other indigenous groups in Guatemala are the Garífuna, with some 5500 people along the Caribbean coast, and a slightly larger population of acculturated Xinka in southeastern Guatemala who are culturally similar to the Maya.

Most Maya live in the hamlets, towns, and urban centers of the central and western highlands of Guatemala, including the Sierra Madres and Cuchumatanes mountains, extending north from the Sierra de las Minas mountains into the southern Petén. The central Petén lowlands have been largely unoccupied by Maya since Classic times.

The K'iche' are the largest indigenous group in Guatemala and Central America, accounting for roughly one-third of the Maya. Mam speakers (700,000) are the second largest, living west of the K'iche' in

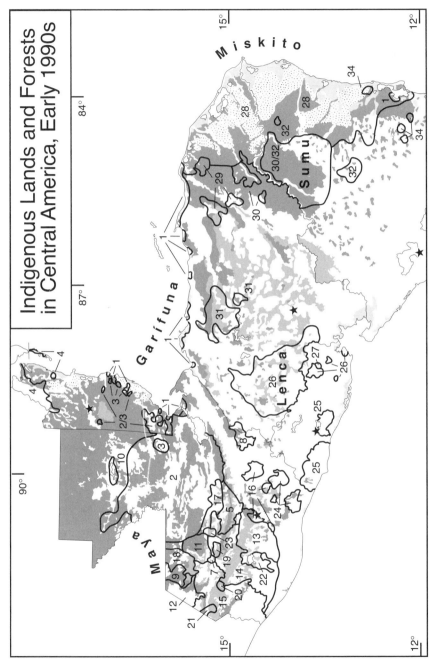

Indigenous Lands and Forests
in Central America, Early 1990s

8–1. Indigenous lands and forests in Central America in the early 1990s.

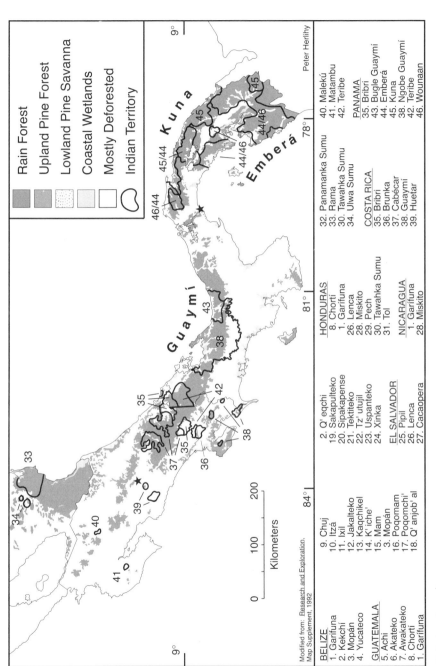

Modified from: Research and Exploration.
Map Supplement, 1992

Peter Herlihy

Rain Forest

Upland Pine Forest

Lowland Pine Savanna

Coastal Wetlands

Mostly Deforested

Indian Territory

K u n a

E m b e r á

G u a y m í

BELIZE
1. Garífuna
2. Kekchí
3. Mopán
4. Yucateco
GUATEMALA
5. Achí
6. Akateko
7. Awakateko
8. Chortí
1. Garífuna

9. Chuj
10. Itzá
11. Ixil
12. Jakalteko
13. Kaqchikel
14. K'iche'
15. Mam
3. Mopán
16. Poqomam
17. Poqomchi'
18. Q'anjob'al

2. Q'eqchi
19. Sakapulteko
20. Sipakapense
21. Tektiteko
22. Tz'utujil
23. Uspanteko
24. Xinka
EL SALVADOR
25. Pipil
26. Lenca
27. Cacaopera

HONDURAS
8. Chortí
1. Garífuna
26. Lenca
28. Miskito
29. Pech
30. Tawahka Sumu
31. Tol

NICARAGUA
1. Garífuna
28. Miskito

32. Panamanka Sumu
33. Rama
30. Tawahka Sumu
34. Ulwa Sumu

COSTA RICA
35. Bribri
36. Brunka
37. Cabécar
38. Guaymí
39. Huetar

40. Malekú
41. Matambu
42. Teribe

PANAMA
35. Bribri
43. Bugle Guaymí
44. Emberá
45. Kuna
38. Ngobe Guaymí
42. Teribe
46. Wounaan

8–1. (continued)

Indian Peoples and Lands Today 219

the forested volcanic highland near the Mexican border. The Cakchiquel (400,000) and the Q'eqchi (350,000) are the country's third and fourth largest groups. Cakchiquel lands lie just west of the capital, focusing on Chimaltenango, Sololá, and Quiché (the town where the famous battle with Alvarado's forces was fought). Many Indian towns are now tourist attractions. The Q'eqchi (Kekchí) live in a wide belt across central Guatemala in Alta Verapaz and Izabal departments but have also expanded north into the Petén and Belize. The remaining seventeen groups are found among these larger populations and account for about 20 percent of the total.

The K'iche' experience exemplifies what Maya groups have confronted in recent times. The nineteenth-century coffee market used K'iche' cultivation of the export crop and wage labor on large estates. Maya scholars George Lovell and Christopher Lutz explain: "A mindset emerged intent on creating a 'coffee state' by deploying native lands and hands to the ends of export agriculture. In the next few decades, the K'iche' lost prized land to aggressive planters, who also established large estates in the less fertile highlands to corral a work force that could be diverted, when needed, to the plantations. Forced-labor laws made up for any slack in this system."

The most troublesome aspects of this political economy were dismantled by 1954, but hopes for progress and land reform disappeared with the military coup in that year. Since then, mostly military regimes have ruled Guatemala, and the Maya have suffered. Demands on their highland lands have been great, and much of the best land is given over to the cultivation of export or cash crops, especially coffee. The pine forests are cleared to meet demands for new agricultural plots and for firewood. This practice coupled with population growth means that the Maya struggle to survive on ever-smaller parcels of land. The K'iche' and other Maya groups are thus at the center of an impending ecological crisis: their populations are growing while resources are declining.

Armed rebellions during the 1980s worsened life for highland Maya. Government forces carried out counterinsurgency campaigns on indigenous areas, especially from 1982 to 1984. The government forces often did not distinguish between the guerrillas and local indigenous peoples, so villagers were massacred, causing massive migrations to Mexico (see chapter 7). Others found haven working on coastal plantations. The Maya repatriation has begun, but the government has been leery of politically organized refugees, and continued land shortages and poor ecological and social conditions threaten progress. Many Mayans still live outside their homelands, and until recently have been afraid to return home. The signing of the peace treaty between the government of Guatemala and the four principal Guatemalan guerrilla groups in 1996 is a major step forward politically for the indigenous peoples. Their population growth, however, puts them at odds with their natural environment, and additional land loss will aggravate the situation.

Guatemala has no reserves for indigenous peoples, but the constitution recognizes that Indian communities hold different cooperative or communal land use systems that warrant special consideration. In theory, indigenous peoples have the same constitutional rights to own land as other citizens, but in practice, because thousands have been killed and hundreds of thousands dislocated during the 1980s, Guatemala's Indians have had little negotiating power with government officials over territorial rights. Their lands are ever more rapidly being invaded and overtaken by outsiders.

El Salvador

El Salvador's indigenous peoples were reduced to landless peasants in the middle of the last century. Most groups are "deindianized" and have lost their native language, dress, and identity. Claims that El Salvador has more than 500,000 so-called invisible Indians whose existence is commonly denied owing to high levels of acculturation may be exaggerated, but much of the confusion is due to unsatisfactory definitions of what an Indian in El Salvador is.

Most surviving Indian heritage is in the west of the country, where Pipil descendants live. Pipils are of Nahuatl-Mexican origin and moved into the area in pre-Columbian times, although they retain few trappings of indigenous culture. They also occupied adjacent Guatemala, but there they are assimilated and barely visible now. In 1932, following the collapse of the coffee market and the world depression, an uprising of Indians led by Agustín Farabundo Martí and others resulted in the occupation of several towns in El Salvador's coffee-growing zone. As a result, about 30,000 Indians and peasants were killed by government troops; facing severe repression, the Indians have abandoned their language and traditions. The country's recent civil war has worsened the situation, leaving many in chronic poverty. Recently, however, the National Salvadoran Indigenous Association has been formed to reassert indigenous rights and bring in development initiatives.

Lenca Indians have also settled in El Salvador: one group is in the northeast, the other along the Gulf of Fonseca, where they are separated from the main population in Honduras. There were about 5000 in the early 1980s, but the recent civil war has further disrupted their populations.

Honduras

The line separating Mesoamerican from South American traditions passes north-south through Honduras, home of seven indigenous groups. To the west, groups are usually merged into the mestizo population; to the east, groups maintain more indigenous traditions.

The Lenca and Chortí once occupied much of western Honduras in settlements bordering the Guatemalan Maya. Today, although both groups maintain certain indigenous cultural and racial distinctions, they have blended with Honduran life and have all but lost their native languages. About 2000 Chortí live around the famous Classic Maya site of Copán in Honduras; they are an eastward extension of the larger Guatemalan population of 50,000. Only four villages contain Chortí speakers, while the rest are without native language, dress, and other cultural traits.

Lenca towns and hamlets, including a total population of 50,000 to 80,000 inhabitants, dominate the landscape of Honduras's southwestern highlands. Conquest and colonization brought them into the Hispanic economy and converted their pine-forested homelands to agriculture. Lenca now have a mixed economy of farming and cattle herding. Their culture can be recognized by Lenca women's dress, pottery, and festivals, but their language may be lost. They recently formed the National Indigenous Organization of Honduran Lenca to campaign at the national level for their land rights. A major march on and demonstration in the capital in 1994 led to the granting of territorial status to one of the large Lenca provinces. The Tol (Tolupanes or Jicaque) live mostly in Yoro Department of north central Honduras in a territory that has been reduced by 70 percent since Spanish contact; it covers 1600 square miles today. Most of the Tol have lost their native traditions and language except around the Montaña de la Flor Reservation. They number more than 18,000 today and have recently set up the Federation of Xicaque Tribes of Yoro to gain land rights and other social benefits.

Eastern Honduras contains groups with South American traits that have for the most part remained isolated from mainstream Honduran life, although the Pech (Paya), who once occupied an extensive area between the Aguán and Patuca Rivers, were enslaved, missionized, and exploited during the colonial period. More recently, expansion of penetration roads has fragmented their lands into islands within a sea of colonists. About 2000 Pech survive today in about a dozen villages in the upland pine and hardwood forests of eastern Olancho, with minor settlement on the Río Plátano in Mosquitia. They speak Spanish, wear store-bought clothes, work for wages, and have lost not only many of their traditions but also much language use. Still, they formed the Federation of Indigenous Pech Tribes in 1985 to gain territorial control and to initiate development and educational programs. They now recognize their language and cultural traditions as vital for reestablishing their identity. In 1991, nine villages received provisional land guarantees from the Honduran Agricultural Institute, the largest being about 35 square kilometers.

The Miskito continue to be the most expansive and dominant indigenous population of the Mosquitia region (pl. 17A). They profited from seventeenth-century commercial relations with the British to obtain firearms and to put other native peoples under their tribute domain. They expanded from the Cabo Gracias a Dios region to control the Caribbean slopes from the Río Tinto in Honduras to the Río San Juan lowlands in Costa Rica. Their territory extended inland along streams in the pine savannas and rain forests, where they farmed rich alluvial soils, hunted, fished, and collected forest products. Miskito men worked in extractive enterprises, gathering forest resins and gums, cutting timber, and harvesting turtles, shrimp, and lobsters. Their coastal society is matrilocal because men are often absent offshore. Miskito villages are scattered along the coasts, lagoons, and inland streams from the Río Tinto in Honduras to Pearl Lagoon in Nicaragua. There are today about 140 villages and more than 35,000 Miskito in Honduras.

Miskito leaders opened a dialogue with Honduran agencies to gain legal control and autonomy over their lands. They formed the country's first indigenous federation in 1976, called Unity of the Mosquitia, to fight for Miskito identity and international support. The Miskito and other ethnic groups of the Mosquitia have been involved in a land legalization program designed to help them understand land use and territorial issues; directed by a Honduran nongovernmental organization, this initiative works on issues of health, education, agriculture, natural resource exploitation, and land tenure.

In the most remote rain forests of the Honduran Mosquitia there still live fewer than 1000 Tawahka Indians, a group belonging to the Sumu people. For centuries the Tawahka occupied most of the Patuca Basin, but Spanish missionaries and soldiers penetrated their lands from the headwaters, then Miskito raiders came upriver and caused further relocation and reduction of Tawahka settlements. Tawahka villagers now speak Miskito more than their own tongue; children are taught Spanish in school, speak Miskito in the village, and use Tawahka with the immediate family. About 650 Tawahka villagers exploit some 770 square kilometers of land for subsistence centered on the confluence of the Wampú and Patuca rivers. The Tawahka formed the Indigenous Federation of Honduran Tawahka in 1987 to fight for land and other social rights. They received marginal recognition of protected status for their lands by presidential decree in 1992 and seek Biosphere Reserve status for a 2331-square-kilometer homeland, centered on their land use area and a binational agreement to establish a reserve system extending from the Caribbean south into the rain forests of the Nicaraguan Mosquitia.

The Garífuna live along the north coast. After their banishment to Roatán, they expanded up and down the coast during the nineteenth century. Over 80 percent of Central America's Garífuna live in Honduras, where forty-four villages include at least 80,000 people, although some estimates go as high as 200,000 to 300,000. Garífuna maintain a strong identity, with distinctive language, dance, folklore, songs, and rituals (see pl. 17B). The women grow bitter manioc as a staple, the only group to do so in Central America today. The men work in offshore fishing or in wage labor, and commercial activities have now brought them into mainstream Honduran life. Out-migration is common as individuals seek opportunities in Honduran and U.S. cities. They formed a federation in 1977 to negotiate for land rights, education, and other needs. Recent mestizo encroachment onto their lands has brought them into the broader debate over indigenous territorial rights.

Honduran law, like Guatemalan, does not include the concept of an indigenous reserve or homeland. A few indigenous communities did receive land titles during the seventeenth and eighteenth century, and twenty-one Tol and two Pech communities gained title during the 1860s. The most recent indigenous land title awarded less than 30 square kilometers to the Tol in 1929—presently the largest legally recognized indigenous area in Honduras. These awards include only small areas around villages, however, and do not reflect indigenous resource-use patterns. The National Agrarian Institute has recently approved about a dozen provisional land guarantees for Pech, Miskito, and Tawahka villages.

Indian cultures are mostly lost to Spanish influence along Nicaragua's Pacific coastal zone. Little remains of the Nicarao conquered by Francisco Fernández de Córdoba in 1524. The Matagalpa fared somewhat better but are now barely distinguishable from mestizo society.

Four indigenous groups survive on the Caribbean side of Nicaragua. The Sumu were the most widespread aboriginal population, including up to ten linguistic groups that lived throughout the Mosquitia's rain forests. They were isolated from sustained relations with the Spaniards, but the dominant Miskito peoples enslaved them and demanded tribute. Only three linguistic groups survive today, the Panamanka (about 70 percent of the total), the Tawahka (20 percent), and the Ulwa (10 percent). Most Tawahka are still primarily forest farmers who engage in cash-earning activities when opportunities arise. Their federation, Sumu Brotherhood, was formed in 1974 to address political, cultural, economic, and territorial concerns.

The Rama are a group of about 1000 Indians who live south of Bluefields Lagoon, and there are two outlying Garífuna villages of about 1000 people on the west side of Pearl Lagoon. They settled there at the end of the nineteenth century and have since lost much of their language and tradition.

Miskito settlements extend from the Honduran border south to the Pearl Lagoon area, including 80,000 to 150,000 inhabitants. Miskito lands have long held privileged status. A reserve was established in 1860 (the only such area ever established in the country) after the British abandoned their protectorate but was incorporated into the Department of Zelaya in 1894. The Miskito attracted attention from international human rights advocates after the Sandinistas came to power in 1979. Miskito militants rejected attempts to incorporate them into the Sandinista socialist state. When the military occupied their villages, a mass exodus began from the conflictive BOSAWAS–Río Coco borderlands. Records show that 16,000 Miskito and 3000 Sumu fled the war zone to seek safety in Honduras. Most of these families had returned by 1990, and Miskito identity is strong today.

The Nicaraguan government, in contrast to former integrationist polities, now recognizes the political force of its indigenous population. The government approved the Autonomy Statute in 1987, which acknowledges indigenous rights to natural resources and guarantees a degree of self-rule in the North Atlantic and South Atlantic Autonomous Regions carved from the former Zelaya Department. The Miskito control the regional assembly of the northern autonomous region, but not that of the southern, which is dominated by Creoles. The autonomy law provides for administration by regional assemblies, but indigenous control hinges on election results. Nicaragua still has no national watchdog agency to oversee indigenous affairs.

Costa Rica

Costa Rica's European heritage contrasts sharply with the few surviving pockets of native life. There were neither masses of Indians nor riches of gold to attract early Spaniards to the Talamancas. The country has about 30,000 indigenous people today, living on twenty-one reserves that cover 3200 square kilometers, all set aside during the past forty years. The reserves are legally recognized by the Indigenous Law of 1977. They are not self-governing, however, and their boundaries are not demarcated or respected by invading colonists or international profiteers, so that they afford little protection for peoples or natural resources.

On the Pacific side of Costa Rica only vestiges of the indigenous peoples survive. About 800 Matambú, once part of the aboriginal Chorotega, live in the center of the Nicoya Peninsula, where they have lost native traditions and language. About the same number of acculturated Quitirrisi and Zapatón live in the highlands south of San José. They no longer speak their Huetar language and are well known to tourists and bargain-shoppers as roadside vendors of pottery and basketry.

Guaymí Indians crossed into Costa Rica from Panama during the early 1900s. About 4500 occupy four reserves south of the Talamanca Highlands in the Coto Brus Valley, on the Osa Peninsula, and around Golfo Dulce. These Guaymí speak Spanish and interact with the national society but maintain their language, customs, and traditions. Long considered foreign migrants, they formed the Ngobegue (People) Association, which successfully demanded recognition of their Costa Rican citizenship. They now struggle with other political, social, and land rights issues. On the southern Talamanca slopes about 2700 Brunka (Boruca) and 1500 Teribe Indians live in three small adjoining reserves. They too are highly acculturated and use Spanish as their principal language but consider themselves Indian.

The Talamancan slopes are also home to Costa Rica's two largest indigenous groups, the Bribri and Cabécar. Having closely related languages, these so-called Talamancan Indians have a combined population of about 17,000. More than 3000 Bribri live in reserves along the Pacific slope, while the remainder shares a reserve with Cabécar on the Caribbean slope. Another small population of 200 Bribri live on the north coast near the Panamanian border. About 8300 Cabécar also live in reserves covering the forests north and south of the Talamancan divide, where their language is reportedly still spoken. Talamanca groups, especially those to the south, have lost much of their land to peasant farmers and cattle ranchers and elsewhere are threatened by large international companies. The Talamancans are organizing to strengthen their opposition to these forces. The Codebriwak Association (Committee for the defense of the rights of the Bribri and Cabécar) unites the efforts of local groups.

Only 520 Maleku (Guatuso) survive on the plains north of the Cordillera de Guanacaste. They survived colonial contact isolated in the San Juan rain forests but were disrupted by rubber tappers, agricultural colonists, and cattle ranchers, who have taken their lands. They may be the most vulnerable native group in Central America. Although many still speak Maleku, they no longer build traditional houses, maintain little of their material culture, and have lost most reserve lands. They now negotiate to regain territory and to reestablish their cultural identity through bilingual education.

In 1973, the National Commission for Indigenous Affairs was set up as an oversight organization for the rights and interests of the country's indigenous population. Indigenous people demand involvement in programs affecting their lands and peoples and are developing local councils to deal with outside pressures. Nevertheless, agricultural colonists have made serious inroads, expropriating lands within the indigenous reserves, in spite of the Indigenous Law, which does not allow these lands to be transferred to non-Indians. Some estimates suggest that as little as 30 percent of reserve lands remain under indigenous control.

Panama

Panama has seven indigenous groups—three east of the canal and four west. Although composing only 8 percent of the population, Indians occupy large territories, where they remained isolated from much outside influence until recent penetration by colonists was triggered by road construction. Panama has a long-standing though only partially implemented tradition of recognizing indigenous homelands.

The eastern interior of Darién was Kuna territory until the eighteenth century, when the Kuna began fleeing from Spanish troops moving across the continental divide to the San Blas Coast. Most of the country's 50,000 Kuna now live in sixty villages on the San Blas Islands and coast. They grow subsistence crops in slash-and-burn fields on the mainland and tend coconut groves on outer islands for cash sale. Kuna women sew colorful tapestries called *molas* for a vigorous market of tourists who visit their islands or buy them in Panama City (pl. 18A).

Some mainland Kuna continue to occupy the interior isthmian rain forests. A few villages occur around Lake Bayano, where the Kuna seek territorial status for their land claim, called Madungandi; five villages, with about 1500 people, remain in Darién.

Kuna identity is probably stronger than that of any other Indian group in Central America. Traditional leaders fit into a highly stratified political organization that includes chiefs (*caciques*), congresses, village leaders, and legislators who together administer their semiautonomous homeland (comarca). The Kuna gained semiautonomous status for their Comarca in the San Blas coastal region (2,357 square kilometers) in 1938. Comarcas are Indian homelands in which there is internal administration of laws and social policies under the jurisdiction of the federal government. The concept emerged during the 1960s with the political support of the former military dictator, Gen. Omar Torrijos. *Kuna Yala,* as the Comarca San Blas is called, is now an institutionalized political force with its own federal legislators, representatives, and lobbyists. As a result, Indian politics are taken more seriously in Panama than in any other Central American country.

Panama's other indigenous groups have now adapted the political and social model of the Kuna Indians for their own organizations. Since the 1960s, Kuna advisors have taught other groups how to structure their political organization. All groups are now struggling to protect their lands from outsiders.

The Emberá and Wounaan, formerly known collectively as Chocó, live in the rain forests of Darién, eastern Panama (pl. 18B). Emberá from the Colombian Chocó region made advances into Darién during the late eighteenth century, after most Kuna had abandoned the area. The Wounaan followed North American missionaries there from the adjacent Chocó during the mid 1900s. Over 15,000 Emberá and 3000 Wounaan now live in Panama. Most are in Darién, where nearly 11,000 Emberá and 2400 Wounaan occupy eighty villages along rivers draining into the Gulf of San Miguel. Others live in central Panama along rivers emptying into the Gulf of Panama, the canal, and the Caribbean.

The Emberá and Wounaan began a struggle for their lands in the 1950s, when they broke their tradition of living in dispersed households in the forest to cluster their houses around schools and missions, where they now learn Spanish and a Christian way of life. The Emberá and Wounaan began to form villages and political institutions in order to strengthen their territorial control and to gain Comarca status after seeing the political success of the Kuna peoples (pl. 18C). The Comarca Emberá-Wounaan, established in 1983, covers two sectors of Darién; Cémaco (2,880 square kilometers), north of the Chucunaque and Tuira rivers, and Sambú (1,300 square kilometers), in the drainage basin of the Sambú River. The Comarca covers about one-fourth of Darién Province and contains more than 8000 inhabitants.

Western Panama has about 140,000 Ngobe and 4000 Bugle speakers, collectively known as Guaymí, the country's largest indigenous group. The Ngobe occupy the rain-forested slopes north of the Cordillera Central in Bocas del Toro, extending outward into Chiriquí and Veraguas provinces. The Bugle (Bogotá) occupy lands between the Chucara and Calovébora rivers, east of the Valiente Peninsula. Most are slash-burn farmers of maize, bananas, rice, beans, and root crops who sell their surplus production. Guaymí political organization is also being modeled after the Kuna, and they have recently been granted Comarca status.

More than 2200 Teribe (Terraba) live along the Río Teribe near the Costa Rican border, where they also seek Comarca status. A small population of Bribri near the Costa Rican border would also be included within the proposed Teribe Comarca.

Recent Panamanian leaders have shown less concern in protecting indigenous groups and their forests. Environmentalist and Indian activists are presently up in arms over talk of completing the Pan-American Highway through the Darién Gap, an action that threatens the forests and peoples of this extraordinary conservation area. Recent years have witnessed increased tensions over such issues with the state. Indian lands have been colonized by agricultural colonists, and loggers continue to cut their timber. Low-level confrontations have occurred as indigenous groups align with international nongovernmental organizations to confront issues of land tenure and resource exploitation.

The Broken Chain

Indians once ruled the Central American isthmus, and the land was covered by forests. Aboriginal horticulturists then cleared forests for fields and settlements, forming a patchwork of cleared, cropped, and forested lands up and down the isthmus. Natural environments and the cultures in them remained an interconnected ecological chain allowing movement of plants, animals, and peoples among habitats, cultures, and continents.

Spanish colonization began to separate the links. Spaniards focused on the Pacific slopes and central highlands, where they enslaved indigenous peoples and forced them to work on estates and in mines. From south to north, such groups as the Cueva, Dorasque, Chorotega, Nicarao, Matagalpa, and Pipil were either exterminated or assimilated into the dominant Spanish and mestizo society. As the bond between their culture and their environment was broken, indigenous peoples gradually lost their lands, their languages, and their cultural traditions.

Nevertheless, a surprising amount of Central America's natural and cultural heritage survived into the twentieth century. The Maya, numbering more than 3 million people, have the greatest diversity, with more than twenty language groups. Outside the Maya and remnant Pacific group areas, practically all other indigenous peoples in Central America have survived through isolation from the mestizo culture in the tropical forests of the Caribbean side of the isthmus and in the Darién. Here, forested habitats and indigenous cultures were largely intact until the mid–twentieth century.

The development of the Pan-American Highway, a continuous road link to eastern Panama, has been the most significant factor in changing this pattern. An integrated construction effort strongly supported by the United States built gravel or paved roads between countries and into their interiors. By the mid 1960s, the highway was largely completed down to central Panama. Ten years later, it was cut into the heart of the Darién as far as Yaviza. Discussions about completing the highway to Colombia are again under way. There is no doubt that the Darién Gap has protected one of the richest natural and cultural regions in Central America. Conservationists, both national and international, indigenous peoples, and many government officials will certainly unite to oppose strenuously any project that threatens such an extraordinary natural resource.

The dramatic demographic expansion between 1940 and 1990, when the population more than tripled to 29 million people (see table 8–1), has also powerfully transformed the lives of indigenous peoples (see chapter 7). Unable to cope with population pressure in the traditional colonial settlements, states allowed numerous roads to be hived off the Pan-American Highway into the forests. Colonization of the jungle has been the solution for righting demographic imbalances and eco-

nomic inequities. International development dollars have helped finance roads that penetrate the Petén, Mosquitia, the Caribbean coastal
lowlands, and Darién. Thus began the nonsustainable cycle of lumbering, slash-and-burn plots, overgrazing, and soil loss that was documented in chapter 7, all of which ends in the push for more forest that
starts the cycle again.

As a result, today, the "chain" of forests and indigenous peoples is
broken. Most of the remaining forested "links" in Central America are
home to indigenous peoples who have experienced limited mestizo
contact. Though some population shifts have occurred, these indigenous peoples have stewarded these lands for centuries without destroying their resources. However, state governments still often consider the remaining Caribbean forests to be frontiers or so-called
national lands ripe for development and give little thought to the rights
of the resident indigenous peoples. Twenty native groups totaling about
1 million people survive in these Caribbean and Darién forests.

Indigenous Peoples and Protected Areas

Very little of Central America is set aside in reserve lands for its indigenous populations. Even the minor reserves set up account for little
combined territory and afford less protection. Today, indigenous federations reject the notion of their lands being set aside as reserves. To
them, reserves convey a pejorative image with a strong sense of racial
and historic paternalism, implying areas of isolation and marginality
dominated by the state. They do not provide for indigenous land tenure,
autonomy, or resource management.

Central American governments have reacted to concerns over accelerated deforestation by trying to save as much as possible of what remains as protected areas. Some 240 protected areas were established in
Central America by 1990, covering about 13 percent of the region, and
more parks are proposed. The protected areas were established, however, with little regard for resident native peoples. They were designed
primarily to conserve nature, but they include the traditional subsistence zones of resident groups. Protected areas contain villages and traditional economic networks that have operated for centuries without
destroying the natural habitats on which they depend.

About 75 of Central America's protected areas are occupied or ex-
ploited by indigenous peoples (fig. 8–2). For example, the Río Plátano
Biosphere Reserve, covering 5000 square kilometers of the Honduran
Mosquitia region, includes Miskito, Garífuna, and Pech settlements.

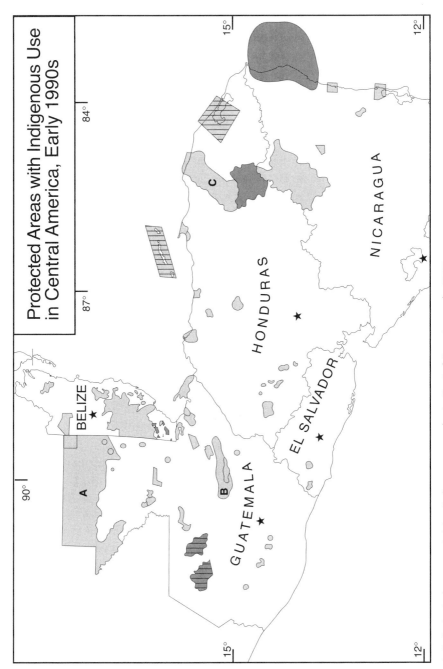

8–2. Protected areas containing indigenous peoples in Central America, early 1990s.

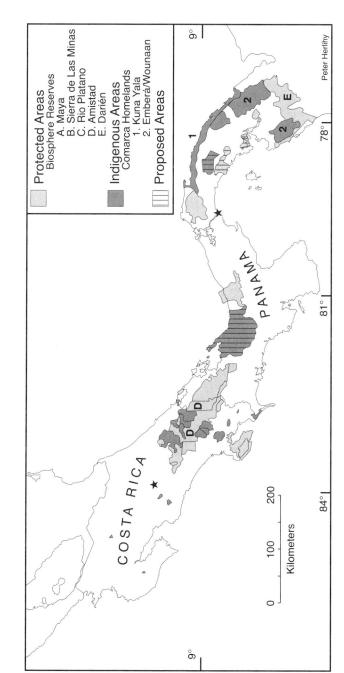

Protected Areas
Biosphere Reserves
  A. Maya
  B. Sierra de Las Minas
  C. Río Plátano
  D. Amistad
  E. Darién

Indigenous Areas
Comarca Homelands
  1. Kuna Yala
  2. Emberá/Wounaan

Proposed Areas

Peter Herlihy

COSTA RICA

PANAMA

Kilometers

0   100   200

9°

9°

84°   81°   78°

8–2. (continued)

Nicaragua's new BOSAWAS Natural Resource Reserve covers 8000 square kilometers of the Río Coco borderlands containing Sumu and Miskito districts. The 6000 square kilometers of La Amistad Biosphere Reserve in eastern Costa Rica contain Cabécar and Bribri lands. Farther southeast, the Darién National Park is the domain of the Emberá, Wounaan, and Kuna.

Although the size and settlement density of native peoples vary from one park to another, their impact is usually sustainable because they are never settled throughout the forest. For example, in 1987, the Comarca Emberá had 8000 Emberá and Wounaan speakers in 35 villages along river margins. Although one of the region's more densely settled rain forests, only half of its land area was exploited, 11 percent being used for agriculture and 40 percent for hunting, fishing, and collecting.

Certain types of protected areas do recognize the need to plan for sustaining both the biodiversity of the forest and the needs of the resident populations. There are two models in Central America.

Biosphere reserves were created to protect natural habitats and their genetic riches while acknowledging the needs and traditions of resident populations. The five biosphere reserves in Central America have all been established since 1980 and are among the region's largest protected areas, covering about half the total parklands (table 8–2). They contain more than 30,000 residents. In practice, however, there has been little regard for native land use, and the reserves presently lack management plans that incorporate indigenous people into the conservation formula. None of the reserves have been able to prevent mestizo colonization and other outside pressures.

The other model for combined conservation and sustainable human land use is the Comarca homelands of Panama. Under the semiautonomous political organization of the Comarcas, indigenous peoples accommodate certain state interests with respect to sovereignty, security, and resource exploitation while retaining authority over their internal cultural, economic, and political affairs. Ideally, the residents should be entrusted with the management of endangered environments while retaining access to resources needed for their cultural and economic well-being. Under the Comarca system, indigenous people have responsibility for managing natural resources in collaboration with state agencies. The Comarcas are set up with regulatory documents that begin to address such management issues. Although land use conflicts and political opposition still threaten the Comarcas, outside invasions are not tolerated, and, in Panama, the Comarca leadership is steadily gaining experience and increased political influence.

Table 8-2. Biosphere Reserves and Indigenous Peoples—Early 1990s.

Biosphere	Year Established	Area (Km²)	Population
Río Plátano	1980	5,251 km²	4,500 Miskito, 250 Pech, and 370 Garífuna[1]
La Amistad	1982	6,227 km²	8,000 Cabécar, 6,500 Bribri, and few Teribe and Guaymí[2]
Darién	1983	5,790 km²	2,039 Emberá, 467 Kuna, and 39 Wounaan[3]
Maya	1990	15,000 km²	unknown native and immigrant Maya
Sierra de Las Minas	1990	8,000 km²	7,500 Q' eqchi Maya[4]
Total		40,268 km²	29,665

1. Herlihy 1991, fieldwork; 2. Luis Tenorio 1988, *Reservas Indígenas;* 3. Proyecto Regional Oriental 1993 (Congreso Emberá-Wounaan-Kuna); 4. Michael Castellon 1994, personal communication.

Indigenous Peoples and the Environment

Indigenous peoples' relation to the land should be a central issue of conservation and development strategies on the Caribbean slope. Indigenous populations are clearly less destructive of natural resources and less heavily involved in market production than other cultural groups; they are users, not buyers, of land, and the connection between their communal land use systems and the conservation of forest resources has been demonstrated. Few hold legal titles to their land, but they recognize land rights inherited through descent lines. Plots that are cleared, planted, or fallowed become the property of the individual who clears them and can then be passed onto children. Land resources are not commodities bought, sold, and exchanged for profit by native groups. Most indigenous groups recognize user rights to agricultural lands and share communal hunting, fishing, and collecting territories that overlap with neighboring communities. Thus, their resource-use areas are normally much larger than their agricultural land use.

Indigenous peoples can deplete or mismanage their resources. Studies of lake cores suggest that prehistoric populations around Lake La Yeguada heavily taxed agricultural lands with slash-and-burn techniques around 2000 B.C. Similarly, studies show that contemporary horticultural societies can overexploit fish, game, and plant resources, especially if they enter into cash economies. Population growth places increasing pressures on native lands. Emberá and Wounaan families of the Darién Biosphere Reserve and Comarca Emberá once moved their settlements when game became scarce but can no longer do so. They have all but eliminated important game and plant resources surrounding the larger settlements. The Miskito have worked in extractive economies that have severely taxed ocean resources, including green turtles and lobsters. At present rates, resident populations in protected areas will double over the next two decades. If indigenous peoples are not given new management skills, natural resources will be destroyed entirely through internal pressures from population growth.

In some cases help has come from missionaries and government programs that have taught Spanish and other Western skills to indigenous groups to improve their ability to deal with the state and with outsiders. Other changes might be attributed to the influence of social scientists who brought indigenous peoples to consider their relations to outside societies. Often, philanthropic sympathizers have helped them to confront social problems and land security issues. Accelerated indigenous activism has also sprung from involvement in regional political turmoil, such as the insurgency movements in Guatemala and El Salvador and the Nicaraguan Contra-Sandinista war. The examples set by the Miskito and Kuna have provided other indigenous activists with inspirational visions.

Nearly all indigenous groups now have a federation or centralized nongovernmental organization representing them, and these have become their de facto government. They are not, for the most part, separatists opposing the states. Most federations seek legally protected homelands with semiautonomous political status. Regional indigenous leaders have also joined international human rights and conservation organizations to defend their lands and cultural identity. The first international conference of Central American Indians was held in Panama in 1977, at which a Regional Council of Indigenous People of Central America was created.

Today, there is a resurgence of indigenous cultural identity that has produced widespread resistance to unfavorable development programs. Indigenous populations now rally around issues of land tenure and natural resource conservation. In 1989, the First Inter-American Indigenous Congress on Natural Resources, sponsored by Kuna and international nongovernmental organizations, was held in Panama. National level congresses on natural resources have followed among the indigenous populations of the Mosquitia in 1992 and of Darién in 1993. These two recent events culminated participatory mapping projects in which indigenous peoples worked with researchers to produce standard maps of their settlements and land use. These activities empower the indigenous populations to negotiate more effectively with government agencies.

Gradually a consensus is emerging that indigenous peoples must be part of the biodiversity conservation formula. Conservationists and government officials have begun to recognize the potential of management strategies that acknowledge the connection between the conservation of natural resources and the recognition of lands and subsistence traditions of indigenous populations. The connection between lack of land tenure, which allows colonists to move onto indigenous lands, and deforestation is increasingly clear. Providing native peoples with legal titles to their communal land has not yet become an acceptable strategy to safeguard regional forest resources. But Indian stewardship has maintained forests on the Caribbean lowlands, and evidence increasingly shows that their communal land ownership promotes the long-term conservation of these resources.

Cultural Identity and Western Conservation Goals

Assertion of indigenous identity in Central America is no longer generally met with integrationist policies. Still, the incorporation of indigenous peoples into national life will likely continue as they are pushed off their lands. Indian life will also fade wherever transportation, communication, and economic networks homogenize life. Much of the physical isolation that formerly sheltered the forests and peoples has been removed.

Most Central American countries do not have defined policies for dealing with their indigenous populations. Native peoples have culture histories and roots that predate international boundaries. Some activists consider their struggle for autonomy within the context of their historic "nation," and some federations have developed separatist ideologies that alienate them from productive state relations. Indigenous issues, however, differ from state to state. Most indigenous groups recognize the specific political and social realities of the countries in which they live. The Honduran Miskito have a separate background from those in Nicaragua, and they did not fight, except when forced to, in the bloody Contra-Sandinista war. The Emberá and Wounaan of Panama have equally different traditions and historic realities from their relations in Colombia. The same can be said for the Maya of Belize and Guatemala; the Lenca of Honduras and El Salvador; the Guaymí, Bribri, and Teribe of Costa Rica and those of Panama. Indigenous issues should increasingly be addressed within the context of population distributions, settlement histories, and political circumstances.

Indigenous identity is on the rise in Central America. Native people are redefining their own "indianness" and demanding recognition of their identity and political institutions. State policies are beginning to acknowledge indigenous authority. Panama has comarcas, Costa Rica has reserves and an Indigenous Law, Nicaragua passed the Autonomy Statute, and Honduras protected indigenous territories for the first time in its history.

Indigenous peoples occupy the bulk of the remaining forests of Central America. However, governments, prompted by conservationists, have established protected areas on their lands, largely without their consent or advice. Thus, the conservation of endangered habitats and the lifeways of the indigenous populations are linked. Miskito, Pech, and Garífuna land uses are core to management considerations for the Río Plátano Biosphere Reserve. The Tawahka Sumu propose their indigenously managed biosphere as the core of Central America's largest rain forest corridor. Conservation laws generally recognize native peoples' rights to well-being on their own lands, and no one considers relocating them off their lands. Yet, conflicts can arise over state desires for conservation or preservation of natural resource and indigenous land use patterns.

Ongoing deforestation and agricultural colonization will almost certainly mean that protected areas will ultimately contain most of the remaining primary forests and much of the indigenous cultures in Central America. Costa Rica, for example, now has little forest outside the same protected areas that contain most of the nation's surviving indigenous peoples. Maintenance of the native farmers and forests of the Caribbean slope and Darién will require the careful management of these large cultural parks. Indigenous peoples have never really been excluded from working with government resource agencies to manage their lands, but this has not normally been the approach. It is now necessary for indigenous and state governments to collaborate for the long-term protection, conservation, and well-being of the region's indigenous peoples and the natural resource base. Doing so may produce a productive relationship to help avoid the regional "ethnic polarization" and subsequent violence predicted by some.

Protected area laws recognize, albeit vaguely, that resident indigenous populations should be given access to the natural resources needed for their well-being. Indigenous land tenure systems, with transfer and use regulated by kin ties, however, afford little protection against land-grabbing outsiders, and an urgent need exists for new land tenure policies toward indigenous lands. Conservation areas appear to offer possibilities, but they do not specifically explain how to deal with the land rights of the resident populations.

The most immediate threat to surviving peoples and forests of Central America is land loss to agricultural colonists. Indigenous lands and protected areas, except in a couple of rare instances, are not delimited or demarcated. Lawmakers need to recognize the communal land use traditions of indigenous populations. What appears at issue are state policies that allocate small plots on a farmer by farmer basis. Agrarian policy is based largely on the notion of individual landownership and does not account for communal land use traditions of indigenous peoples. Indian settlements have overlapping spheres of resource use in which different villages hunt, fish, collect, and farm on the same territories. New land tenure approaches advocate viewing traditional land use from an interrelated, communal, or regional perspective. Land legalization issues must address the rights at various territorial scales of the community. Unless protected areas management allows indigenous peoples to govern themselves by their own cultural traditions, there is probably not much hope.

Indigenous peoples in Central America want development on their lands. With growing pressures on their resource bases, many communities now lack clean drinking water and need sewage disposal systems. Villagers also want cash to meet consumer needs. Native groups now look for development alternatives to provide economic possibilities without destroying natural resources. Researchers align with them to explore modes of "constructive exploitation" that promote the sustained use, conservation, and preservation of natural habitats while offering economic gain. Ecotourism has been one widely publicized option by which local people can replace incomes generated from resource extraction with that from a sustainable program of tourism. Market-oriented agroforestry and crafts supply many groups with cash supplements that do not require the cutting of the forest. Ecologically sound and economically sustainable strategies can be developed together with the indigenous peoples.

Conservation and human rights nongovernmental organizations put the peoples up front in the formula to manage biodiversity. Yet, beyond rhetoric, few development and conservation projects have much involvement at the local level. Conservation and human rights organizations deal more with a small number of leaders than with communities. Projects and research circle around the political world of the indigenous federations, which the nongovernmental organizations usually finance and influence politically. Dangers lie in the fact that federations may not be representative of what is occurring in the countryside, reflecting more the thinking of a few educated Indians in centralized federation or nongovernmental offices, rather than that of their relatives and masses in the forest. They have a top down organizational structure and rely on information and decisions of a small group of specialists rather than of the grassroots.

The conservation of biodiversity in Central America probably will not work without the involvement of local peoples. New strategies need to incorporate villagers into a formula that makes conservation more profitable than exploitation. Indigenous people can play an important role in the ecotourism model as forest rangers and tour guides, but they could do more if trained as ethnoscientists, agroecologists, and managers of their lands and resources. Governments, nongovernmental organizations, and researchers alike need to develop truly grassroots, participatory approaches toward human rights and conservation issues. Native peoples should be at the center of conservation management. Only with their help might the forested links be rejoined to reestablish the ecological connectivity of Central American environments and to make the Paseo Pantera a reality.

CHAPTER NINE

The Paseo Pantera Agenda for Regional Conservation

JORGE ILLUECA

It was against the evolving backdrop of the Central American peace process and the revival of integration efforts that the Paseo Pantera was first conceived in 1990. It is a bold and ambitious project that proposes to unite protected areas throughout the length of the Central American isthmus so that the necessary movement of wildlife will be achieved by means of ecological corridors (fig. 9–1). The goals are to preserve the region's immense biodiversity and to contribute to the economic recovery of the isthmus by the promotion of sustainable development activities such as ecotourism and agroforestry. Because conservation is increasingly perceived in Central America as an instrument of peace, the Paseo Pantera has appealed to a number of groups in the region. The concept was also favorably received at the IV World Congress of Parks and Protected Areas, held in Caracas, Venezuela, in February 1992.

Toward Peace and Integration

The 1980s, commonly referred to as Latin America's lost decade, was particularly severe in Central America. During these years the region was gripped by a profound political and economic crisis, the economic effects of which have been reviewed in chapter 8. They can be summarized by noting that the per capita gross domestic product of all the Central American countries, excluding Belize, plummeted by an accumulated 18.3 percent from 1980 to 1990 as prices for leading exports declined in international markets; and for most, gross domestic product

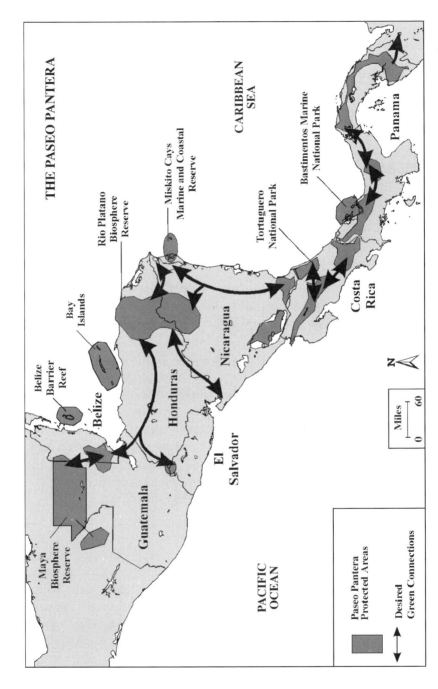

9–1. Map of the major protected areas of the Paseo Pantera and the links between them that would form a complete system of ecological corridors.

levels today remain inferior to those of 1978. For Central American society in general, this translated into declining income, rising unemployment, loss of purchasing power, and a lowering of the quality of life. Trade among the five member countries dropped from US$1.1 billion in 1980 to US$420 million by 1986. Total exports fell from US$4.6 to US$3.8 billion from 1979 to 1987. In spite of the bleak picture, however, the period also witnessed a major expansion of protected areas in Central America

Although serious political differences obstructed the implementation of a peace agenda for Central America, there generally existed agreement among government environmental managers on a cooperative environmental agenda. Conservation, the antithesis of war, became an instrument of peace in Central America.

The end of the Cold War signaled the beginning of the end of armed conflict in Central America. A very important role in the diplomatic solution of the conflict was played by the Contadora Group, consisting of Panama, Mexico, Colombia, and Venezuela. It was largely owing to the group's efforts that the Central American presidents were able to negotiate and adopt the "Procedures for Establishing a Firm and Sustainable Peace in Central America" at the II Presidential Summit of August 1987 in Esquipulas, Guatemala. With the exception of Nicaragua and Panama before the December 1989 invasion, the economies of the region then entered a period of recuperation that to date continues.

The Esquipulas II Declaration affirmed that peace, democracy, and development are inseparable in Central America. Improvements in one are not possible without parallel advancements in the others. Five areas of action were agreed upon: the promotion of democracy and political reconciliation, the termination of all support to insurgent movements, the repatriation and resettlement of refugees, the negotiation of arms reduction and other security agreements, and cooperation in seeking international assistance for the recovery of the region.

Economic reconstruction had as a point of departure a reevaluation of the General Treaty of Economic Integration of 1960, the foundation for the Central American Common Market. The Summit of Central American Presidents in Guatemala in 1990 proposed a reformulation of the concept of economic integration that was oriented more toward greater participation of the region in the world market and less toward protectionism. As the Economic Action Plan has become more refined through implementation, it is clear that development priorities in the new scheme differ from those conceived initially for the Common Market. Greater emphasis has been placed on combating poverty and protecting the environment.

Specifically, the Central American states agreed to the following objectives: (1) strengthening of the protection of the natural heritage; (2) adoption of sustainable development programs; (3) the optimal utilization of natural resources; (4) the control of pollution; and (5) the reestablishment of an ecological equilibrium. To these ends, special attention is to be given to the improvement and harmonization of national environmental legislation and to the funding and implementation of conservation projects.

Nonetheless, despite gains in the gross domestic product, the strengthening of the Central American democracies, and improved regional cooperation since 1988, economic and social instability continue to plague the isthmus. The number of people living below the poverty line in Central America has increased from 67 percent in 1989 to 73 percent in 1994. If environmental programs are to be viable, they must aid in solving the problem of poverty that threatens to destabilize the entire region.

The Building Blocks for Paseo Pantera

The Commission on Environment and Development

The Central American presidents agreed in 1989 to create the Central American Commission on Environment and Development (henceforth the commission). The commission is committed to evaluating and protecting the region's rich biological diversity, including control of illegal trade in wildlife, and to requesting financial and technical support for biodiversity projects. In doing so, it promotes coordinated action among governmental bodies in such areas as management of natural resources, particularly tropical forests, and the protection of watersheds and transboundary ecosystems. It is increasingly playing an important role in capacity building of national environmental institutions. Its members are appointed by their respective governments but generally consist of the ministers or directors in charge of the environment and natural resources. Perhaps its greatest achievement is the regular inclusion of environmental and sustainable development issues in the agendas of the Summits of the Central American Presidents.

Convention for the Conservation of Biodiversity and Protection of Priority Wildlife Areas in Central America

Created on International Environment Day, June 5, 1992, the Convention for the Conservation of Biodiversity and Protection of Priority Wildlife Areas in Central America (henceforth the convention) covers a wide range of biodiversity issues that are important to the Central American governments. The objective of the convention is to protect to the utmost the region's biodiversity for the benefit of present and future generations.

In several areas, the convention proposes actions that are relevant to the Paseo Pantera initiative. It encourages cooperation among countries in border conservation activities, assigning top priority to eleven border protected areas. Most important, it establishes a Central American Council on Protected Areas that is responsible for "the development of a Regional System of Protected Areas as an effective Mesoamerican biological corridor." It also supports the promotion of ecotourism as an important source of income that can generate more employment and provide funds for the operation of protected areas.

The Central American Commission on Environment and Development would be the appropriate body to establish the Paseo Pantera legally, and in cooperation with the Central American Council on Protected Areas, which is comprised of the national directors of parks and protected areas, could oversee its coordination and implementation.

Regional Convention for the Management and Conservation of Natural Forest Ecosystems and the Development of Forest Plantations

At the time of the approval of the convention in 1993, it was estimated that 13 million hectares of Central American forests had been eliminated on lands suitable only for forests. Moreover, it was estimated that 418,000 hectares of forests were being eliminated annually in the region. The convention aims to protect existing forests located on lands classified as suitable only for forests and to restore forests on such lands that have been denuded of their natural vegetation.

Recognizing that approximately 20 million persons in the region live in poverty and that two-thirds of these reside in rural areas, the convention addresses the need to promote the sustainable development of forest resources through national and regional reforestation, plantation and agroforestry programs that take into account local participation and help counteract poverty.

This convention could be an important building block for the Paseo

Pantera. It calls for "the consolidation of a National and Regional System of Protected Wildlands, that ensures the conservation of biodiversity and the maintenance of vital ecological processes." Its objective of restoring forests on degraded lands classified as suitable only for forests could help in restoring the ecological corridors connecting the protected areas that are the basis of the Paseo Pantera.

The Commission on Environment and Development is charged with the lead role in seeking financial support for these efforts and for the establishment of a Central American Council on Forests, consisting of forest service authorities designated by each state. The convention is expected to become effective by 1995.

International System of Protected Areas for Peace (SIAPAZ)

The year 1990 was declared the Year of Peace and Reconstruction by the government of Nicaragua, and in October of that year Costa Rica and Nicaragua jointly established the International System of Protected Areas for Peace (SIAPAZ) and created a high level binational commission to run the program, which was declared to be of the highest priority for both governments.

SIAPAZ is geographically at the heart of the Paseo Pantera. It is an impressive complex of protected areas rich in biological diversity and containing rain forests, rivers, wetlands, and lagoons, extending from Lake Nicaragua down the San Juan River to the Caribbean coast.

In Nicaragua, its jewel is the Indio–Maíz River Biological Reserve containing the least altered forest in the country. Bordering the San Juan River and the Caribbean coast, it has been declared a reserve for the absolute protection of biodiversity and is open only to scientific research and ecotourism. The Guatusos Wildlife Refuge is a strip rich in wetlands between Lake Nicaragua and the Costa Rican border, extending from the mouth of the Pizote River to San Carlos, where the San Juan River exits Lake Nicaragua. Other protected areas of the system include the Solentiname Archipelago National Monument along the archipelago of the same name in the southeastern part of Lake Nicaragua, the Inmaculada Fortress Historical Monument, and the San Juan Delta Wildlife Reserve. These protected areas, including those of limited use, encompass an area of approximately 5300 square kilometers within a larger territory of 12,700 square kilometers.

In Costa Rica, SIAPAZ includes the Caño Negro Wildlife Refuge, an important marsh area with a high concentration of migratory bird species; Barra del Colorado Wildlife Refuge; Tortuguero National Park, renowned as a sanctuary and breeding ground for sea turtles; the Laguna Maquenque and Tamborcito Wetlands; the two-kilometer-wide buffer zone all along the international border; the Maleku Indian Reservation, housing one of the country's smallest Indian groups; and privately managed forests.

For SIAPAZ to succeed, it needs to significantly diminish the problems of poverty in southeastern Nicaragua and northeastern Costa Rica by implementing sustainable development projects. SIAPAZ currently offers important support to the management of protected and limited-use lands, but its role needs to be expanded to include the promotion of environmentally sound, integrated management of natural resources.

Border Development and Integration

Border development and integration is a priority of the Central American peace plan. In general, frontier areas are politically and economically marginalized and poor in infrastructure. They usually have the highest incidence of poverty. At the same time, because of lower population densities, they are often areas dominated by forests and other natural ecosystems. In the past, the combination of marginalization, poverty, and dense vegetation cover have made some of these areas fertile grounds for armed insurgency.

Economic integration of Central American nations as conceived in 1960 focused almost exclusively on commerce generated from the capital cities and their most important satellite centers of production. Despite the fact that national territories within an international area, for example, a watershed, could be dedicated to complementary or similar economic activities, the integration process of the 1960s did not take into account the need for border areas to coordinate production, marketing, and services. Moreover, the sustainable development of an international watershed or forest ecosystem requires cooperation among member countries in order to avoid negative transboundary impacts such as pollution, sedimentation, or the spread of diseases, which can undermine national plans and programs.

By 1993, with the help of the Organization of American States, the Central American governments developed a plan to coordinate development of border areas as part of a larger economic integration initiative. The eighteen projects of the plan have three common objectives: (1) the regional development of border areas, (2) the arresting of environmental degradation and the implementation of environmental land management programs, and (3) the alleviation of poverty. They propose improvements in social and economic infrastructure, social services, local government capabilities, community participation, and private sector activities. Where applicable, they address the resettlement of refugees and the regulation of activities involving migratory workers. Priority is given to the development of tourism, especially ecotourism. All contain environmental management components in areas such as the management of protected areas, reforestation programs, efficient utilization of fuel wood, and environmental education. Those dealing exclusively with the integrated management of transboundary watersheds contain components on soil conservation and forests.

Taken collectively, the eighteen projects of the plan include an impressive array of protected areas the length of the Central American isthmus:

Mexico	Calakmul Reserve (part of Maya Biosphere Reserve)
Belize	Río Bravo Reserve (part of Maya Biosphere Reserve)
	Chiquibul/Maya Mountains Biosphere Reserve
	Chiquibul Special Protection Area
Guatemala	Maya Biosphere Reserve
Honduras	Copán Maya Ruins
	part of the Plapawans protected areas system consisting of the Plátano River Biosphere Reserve, the Tawahka Anthropological Reserve, and the Patuca National Park
Nicaragua	Miskito Cays Marine and Coastal Reserve
	Laguna de Pahara Natural Reserve
	Laguna Bismuna Natural Reserve
	BOSAWAS Nature Reserve
	Río Escalante-Chococente Wildlife Reserve
	Indio-Maíz Biological Reserve
Costa Rica	Guanacaste National Park
	Santa Rosa National Park
	Caño Negro Wildlife Refuge
	Cahuita National Park
	Gandoca-Manzanillo Wildlife Reserve
	Cocles-Kekoldi Indigenous Reserve
Panama	Bastimentos Island Marine National Park
	San Juan River and Ciénega de Changuinola protected wet lands (proposed)
	Darién National Park
Colombia	Los Katios National Park

A particularly important transfrontier project is the La Amistad International Biosphere Reserve. La Amistad Park has 1939 square kilometers in Costa Rica and 2070 in Panama, and with its impressive area of protected and limited-use areas totals 8178 square kilometers.

In general, the Plan of Action for Border Development and Integration is an overlapping, complementary regional initiative to the Paseo Pantera. Practically all of the priority protected border areas of the Central American biodiversity convention are also treated here. Although the plan covers only border areas, its projects are promoting the preparation of management plans for a large number of protected areas and their buffer zones that would form major links along the Paseo Pantera. They generally include the elaboration of land use management schemes based on geographic information systems for the study areas. The land use management maps can be helpful in the designation of areas for reforestation and agroforestry and for establishing some sections of the ecological corridors of the Paseo Pantera.

Alliance for Sustainable Development

In 1992, President Bill Clinton announced that environment would be one of the cornerstones of a revised U.S. policy toward Latin America. In 1994, his administration hosted the Summit of the Americas in Miami, at which environment and sustainable development were addressed.

In preparing for the Summit of the Americas, the Central American governments proposed to the United States that Central America be designated a world pilot zone for sustainable development, and it was recommended that an Alliance for Sustainable Development be established by mutual agreement of the governments of the United States and the Central American Republics, within the framework of the declaration of the United Nations Conference on Environment and Development. The alliance was formally agreed upon by the Central American governments and the United States at the special Ecological Summit in Managua in October 1994.

This modest proposal has become the most important initiative of the Central American governments with the United States, and the new strategy for national and regional development. The alliance has four principal areas of action:

1. A revision, harmonization, and strengthening of national and regional legislation, including the consideration of more effective coordination between Central America and the United States in the enforcement of legislation regulating the transboundary movement of hazardous substances and wastes;
2. Economic and financial projects, especially involving commerce, investments, foreign aid, debt settlement, and technical assistance in evalu-

ating natural resources and developing cleaner technologies, as well as the proposed integration of Central America into the North American Free Trade Association (NAFTA);

3. The development of alternative sources of energy, increased efficiency in production, and a greater commitment by lending agencies to funding sustainable energy projects;

4. The appraisal, conservation, and utilization of biological and cultural diversity.

Through the Alliance for Sustainable Development, Central America is seeking support for the consolidation of a System of Protected Wildlife Areas, both at the national and regional levels, which will ensure the conservation of biodiversity and the maintenance of vital ecological processes. Such a system would of necessity consider the environmentally sound management and sustainable development of lands bordering protected areas as complementary to the management of the latter. The Paseo Pantera should be at the center of this initiative.

The alliance also calls for the creation of a regional network of biodiversity research centers and botanical gardens and a Central American Biological Corridor. It also sets forth a request for technical cooperation for the elaboration of an inventory and diagnostic study of the biodiversity of Central America, which should include support for the training of Central American professionals in taxonomy, conservation, environmental management, and sustainable utilization of biodiversity. A major initiative should be launched for valuing the region's biodiversity from pharmaceutical, industrial, scientific, historical, and aesthetic points of view. It also calls for greater cooperation in controlling more strictly the trade in endangered wildlife.

The Paseo Pantera Agenda

A project of the magnitude of Paseo Pantera will require substantial funding for its planning, implementation, and management phases. The agenda is the restoration of a Central American land bridge to allow the necessary migration of animals in the preservation of the region's wildlife. In large part, the initiative is a response to the fragmentation of the vegetation of the isthmus into isolated pockets and islands whose viability is questionable. Much of the required resources could be provided through the Alliance for Sustainable Development.

The Paseo Pantera project has been oriented toward capacity building of national conservation institutions by providing them with methods, tools, and knowledge. It contained major components in buffer zone management, regional ecotourism, and environmental education. It also promoted the development of transboundary protected areas.

In June 1990, the Paseo Pantera received financial support for five years from United States Agency for International Development (USAID). A $1.6-million matching grant was given to the Wildlife Conservation Society and the Caribbean Conservation Corporation to develop, in collaboration with governments and nongovernmental organizations, pilot management plans for key existing and proposed protected areas. These included the Tikal National Park, noted for its archaeological splendors and stunning biodiversity, located at the heart of the proposed trinational Maya Biosphere Reserve; the proposed Belize Barrier Reef Biosphere Reserve, the largest coral reef in this hemisphere; the Plátano River Biosphere Reserve and the proposed Bay Islands protected areas network in Honduras; an expanded Tortuguero National Park in Costa Rica; and the Bastimentos Marine National Park in Panama. Another fifteen to twenty protected areas were to be added subsequently, focusing on the establishment of buffer zones and ecological corridors linking this system of protected areas.

A multinational project of this extent, however, surely requires the endorsement and active collaboration of the Central American governments and their citizens. If this is to be achieved, a much greater participation of intergovernmental organizations is desirable, beginning with the Central American Commission on Environment and Development as the principal intergovernmental forum in the region for considering cooperative environmental actions. The commission can explore different options and approaches for officially legitimizing the Paseo Pantera. The Organization of American States could be an important intergovernmental organization providing technical cooperation in the design, planning, and implementation of the Paseo Pantera. Not only could it bring to bear its technical expertise on the project, but it also possesses diplomatic skills that would help to promote acceptance of the project among the governments. Support could also be forthcoming from other intergovernmental organizations, such as the United Nations Environment Programme, which has developed a partnership with the Organization of American States in the La Amistad Biosphere Reserve and San Juan River projects, both of which have important biodiversity management components. This would signify a broadening of the coordination and institutional base of the initiative, in which Wildlife Conservation International and the Caribbean Conservation Corporation would be joined and reinforced by intergovernmental organizations as partners in making the Paseo Pantera a reality.

The Configuration of the Paseo Pantera

In 1992, there existed in Central America 162 protected areas with a combined surface area of 54,000 square kilometers and 71 limited use protected areas described as forest reserves, Indian reserves, protective zones, and multiple use zones, which together account for 87,000 square kilometers, or more than 16 percent of all Central American territory. In order to complete a regional system of protected areas that would protect the range of Central American life zones, the World Conservation Union (IUCN) has proposed the addition of another 10,000 to 20,000 square kilometers of protected areas.

Connecting all of these protected areas would be impossible, prohibitively expensive, and probably not the most effective way to spend limited funds on biodiversity conservation. In spatial terms, Central America's protected areas are very fragmented and often isolated from each other. This poses serious problems because reconnecting many of these protected areas would require ecological corridors that in many cases would have to be very long and relatively narrow.

Logically, the trunk of the Paseo Pantera would run along the less disturbed Caribbean side of the isthmus, from the mountains of the continental divide to the coast. It is along this strip that the largest protected areas of the isthmus are found. Five protected areas along this route account for approximately half of the 54,000 square kilometers of protected areas in Central America. These are the Tigre National Park in Guatemala, the Plátano River Biosphere Reserve in Honduras, the BOSAWAS Biosphere Reserve and the Miskito Cays Marine and Coastal Reserve in Nicaragua, and the Darién National Park in Panama. The Miskito Cays Marine and Coastal Reserve is approximately 12,350 square kilometers of reefs.

The most pressured and isolated parks and wildlife reserves are on the drier, more populated Pacific side. Consideration has to be given to branches from the trunk extending across to the Pacific coast, such as to the Santa Rosa, Guanacaste, and Corcovado National Parks in Costa Rica and the Gulf of Fonseca. The Darién National Park, a World Heritage Biosphere at the southeastern extremity of the Central American isthmus, is the only large protected area that crosses from near the Caribbean coast to the Pacific Ocean, although the proposed Interoceanic National Park of the Americas and the Camino de Cruces National Park in Panama in the region of the Panama Canal come very close to linking both coasts.

The Relevant Legislation

The creation of ecological corridors and protected area buffer zones, where the latter do not exist officially, requires a solid understanding of legal dispositions pertaining to land use and ownership. The review, analysis, and harmonization of environmental management and sustainable development legislation is the work of the Central American Commission on Environment and Development. It could take the lead in promoting and developing this activity. To this end, it should establish links with United Nations Environment Programme's Regional Office for Latin America and the Caribbean, headquartered in Mexico City, which possesses the most comprehensive environmental legislation data base in the region.

Attention also has to be given to the consideration of different legal instruments for officially establishing the Paseo Pantera. The Central American Commission for Environment and Development is probably the best forum for carrying out this task. However, the Paseo Pantera could also be established legally through the Convention for the Conservation of Biodiversity and Protection of Priority Wildlife Areas in Central America, in which case it would require formal adoption at a Summit of the Presidents of Central America and subsequent ratification by the national parliaments. Alternatively, the governments could decide to establish the Paseo Pantera through a new convention. These various options could be considered by the commission if it was given more direct participation in phase two of Paseo Pantera.

Sustainable Development Components

For many of the 20 million impoverished people in Central America, the invasion of forested lands, including protected areas, is one of the very few options for survival that they have. A high percentage of the forests that remain today are found on marginal lands that are unfavorable for traditional agricultural production because of poor soils and their high susceptibility to erosion. Lands are colonized, and habitats rich in biodiversity are replaced by subsistence farms, only to be abandoned after a few years because of soil depletion. They are quickly converted to pasturelands by affluent ranchers. Unless viable economic alternatives are made available to the rural poor, this cycle of poverty will continue until the forests of the region have been irretrievably destroyed.

Compatible, environmentally sound development activities in ecotourism, agroforestry, reforestation, sustainable agriculture, sustainable fishing, and sustainable utilization of biodiversity must be structured and organized to provide social and economic benefits to the rural poor, particularly in the proximity of the Paseo Pantera. As indicated earlier, this is largely the approach adopted by the governments in the Plan of Action for Border Development and Integration.

In particular, ecotourism is a promising option. The Paseo Pantera can provide critically needed economic support through the creation of a wide range of jobs, services, and merchandise, including agricultural goods and folk art. It can be structured to incorporate economic incentives in support of conservation as well as to generate income for the management and expansion of protected areas.

Costa Rica is currently the only country that has taken advantage of its ecotourism potential. It is the leading industry in the country followed by bananas and coffee. Guatemala and Panama are just beginning to realize the potential economic benefits.

Once organized and in operation, the Paseo Pantera will rival the Inca Trail in the Andes and the Appalachian Trail in the northeastern United States as one of this hemisphere's most important attractions for ecotourists.

Agroforestry, reforestation, sustainable agriculture, and sustainable utilization of biodiversity are activities that can collectively employ a substantial labor force and reduce the number of rural poor encroaching on forested lands. The promotion of sustainable agriculture throughout the Central American isthmus, particularly in the proximity of forested areas, is essential to holding the agricultural frontier in place. Massive reforestation activities will help to push it back. Where corridors and buffer zones are to be established, the land use management zoning scheme of the Paseo Pantera would be useful in identifying appropriate areas for developing sustainable agriculture and agroforestry and for carrying out reforestation projects.

Protected Areas Management

The existence of a large number of protected areas in Central America can create the mistaken impression that conservation is doing well in the region. Most protected areas, however, do not have management plans, and their boundaries are often poorly defined. The great majority are neglected as a result of limited human and physical resources. Many have no permanent personnel. Excluding Costa Rica, the Central American countries do not have well-coordinated national systems of protected areas.

For the protected area units of the Paseo Pantera to function properly, priority attention has to be given to strengthening their management. This should be done on two levels. At the regional level, attention has to be given to fortifying the capabilities of national institutions responsible for protected areas and providing them with technical assistance. Operational linkages among the components of the national protected areas systems forming the Paseo Pantera have to be established and reinforced. At the local level, the management of specific protected areas has to be greatly improved, starting with the preparation or updating of management plans, the delimitation of boundaries, the development of required infrastructure, and the training of personnel.

Research

Strategies for preserving biodiversity in restricted areas suffer from insufficient knowledge of spatial requirements. Craig L. Shafer, in his *Nature Reserves: Island Theory and Conservation Practice*, wrote, "Of all the biological approaches to conservation, the continuing study of natural area isolates is our most pressing need, and . . . this field of study must be considerably developed before long-range conservation strategies can be well defined." Unfortunately, little time remains, and it may well be possible that by the end of this century few options will be left for reserve planning and organization, especially as concerns size determination.

The capacity of existing national parks and reserves to afford protection to species and ecosystems requires careful examination and consideration. If they are inadequate in size, loss of biodiversity can be predicted. Of the 209 breeding forest bird species found in 1923 on Barro Colorado Island in the Panama Canal, one of the best protected and managed reserves in the Neotropics, nearly 60 disappeared over a little more than half a century. Forest succession explained the disappearance of some of these species, but not all.

The creation of an interconnected system of protected areas may prevent many species from becoming isolated and dwindling into nonviable populations. In any case, the establishment and development of the Paseo Pantera will provide an excellent opportunity for the study of reserve and corridor sizes needed for the survival of different species and habitats. In order to determine optimal sizes for different kinds of wildlife reserves, habitat size and species loss, as well as minimum viable population requirements, need to be studied far more extensively. In order to predict the effects of future land use change, basic information is also needed on the ecology of rare and endangered species.

For management purposes, this research has to be accompanied by studies of the social and economic needs of people living in adjacent areas, with a view to harmonizing their aspirations with the goals of preserving biodiversity.

Education

Subsequent work of the Paseo Pantera must also continue educating national audiences, local communities, and visitors on the uniqueness of Central America's natural wonders and the importance of the region's biodiversity. The education component should also emphasize the social and economic benefits that can be derived from the Paseo Pantera through ecotourism, agroforestry, reforestation, sustainable agriculture in buffer zones, sustainable utilization of biodiversity, and biotechnology. Materials have to be developed that will help win over the support of local communities and groups. National and local nongovernmental organizations should be invited to play a key educational role in enlisting public support for the initiative.

With its establishment, the Paseo Pantera will be one of the most impressive systems of protected areas on earth. It will be the backbone for protecting one of the richest regions on earth for biodiversity. Intrepid explorers who dare to travel its length will have the incredible opportunity to climb the Maya Pyramid of the Jaguar in Tikal, snorkel over the Belize Barrier Reef, explore the rain forests of the Plátano River Biosphere Reserve, swim after sea turtles feeding in the Miskito Cays, canoe up the San Juan River to Lake Nicaragua to view freshwater sharks, peer into the smoldering crater of the Poás volcano, rise through the cloud forests of the La Amistad International Park, scale the Spanish colonial fortifications at Portobelo, and follow the tracks of the panther into the largely unexplored Darién National Park.

The Paseo Pantera will also stand as a monument to the better side of humanity—that which values peace through conservation over war and destruction. For it to succeed, it must also promote sustainable development and contribute to the elimination of poverty.

Editor's Postscript

Since this manuscript was written, there have been several developments in the implementation of the Paseo Pantera by international organizations and the governments of Central America. First, Paseo Pantera has a new formal name: the Mesoamerican Biological Corridor.

Following the signing of the agreement between the United States and Central American governments, the United States committed all of its financial assistance for the environment in Central America to developing and consolidating the Biological Corridor. The initiative is called proarca and will provide about $10 million over the next three years. The Global Environmental Facility, through the United Nations Development Fund, has pledged approximately $14 million toward the Biological Corridor for a three-year period starting in 1998. For the same period, the World Bank has also committed to invest substantially in projects that are related to the development of the Biological Corridor in Honduras, Nicaragua, and Panama. Last, the European Union, working with the government of Germany, has committed approximately $15 million to assist with the Central American Biological Corridor starting in 1997–98.

Contributors

Anthony G. Coates
    Smithsonian Tropical Research
    Institute
    Republic of Panama

Paul Colinvaux
    Smithsonian Tropical Research
    Institute
    Republic of Panama

Richard Cooke
    Smithsonian Tropical Research
    Institute
    Republic of Panama

Luis D'Croz
    University of Panama and
    Smithsonian Tropical Research
    Institute
    Republic of Panama

Stanley Heckadon-Moreno
    Smithsonian Tropical Research
    Institute
    Republic of Panama

Peter H. Herlihy
    Geography Department
    University of Kansas

Jorge Illueca
    United Nations Environment
    Programme
    Nairobi, Kenya

Jeremy B. C. Jackson
    Smithsonian Tropical Research
    Institute
    Republic of Panama

David Rains Wallace
    Freelance writer
    1568 San Lorenzo Ave.
    Berkeley, California

S. David Webb
    Florida Museum of Natural
    History
    University of Florida

Further Readings

Chapter One. The Forging of Central America

Plate Tectonics

Cox, A., and R. B. Hart. 1986. *Plate Tectonics: How It Works.* Blackwell Scientific Publications, Oxford and New York.
An extremely clear and superbly illustrated account of the mechanisms of plate tectonics. The book contains substantial mathematical calculations and goes into considerable detail. Suitable for an interested reader with a good mathematical background but may be browsed fruitfully by someone without quantitative skills.

Levin, H. 1978. *The Earth through Time.* W. B. Saunders, Philadelphia and London.
A good, extensively color-illustrated introductory textbook of geology with a full account of plate tectonics for the lay reader.

McPhee, J. 1983. *In Suspect Terrain.* Farrar, Straus, and Giroux, New York.
———. 1993. *Assembling California.* Farrar, Straus, and Giroux, New York.
Two wonderful pieces of literature that incorporate lucid accounts of many aspects of plate tectonics while describing different regions of the United States. In each book, the author describes his understanding of the geology and local history as he travels with professional geologists.

The Geology of Central America

Dengo, G., and J. Case (Editors). 1990. *The Geology of North America.* Volume H. *The Caribbean Region.* Geological Society of America, Boulder, Colorado.
The most comprehensive technical account of the geology of Central America and the Caribbean region by many of the leading specialists in the field. The

reader would need some background in geology. For students of geology this is a goldmine of information, including a comprehensive bibliography of previous research.

Jackson, J. B. C., A. Budd, and A. Coates (Editors). 1996. *Evolution and Environment in Tropical America*. University of Chicago Press, Chicago.
This book contains chapters by several authors concerning the rise and closure of the Central American isthmus during the last 10 million years, and their biological consequences. It includes a fairly detailed account of the regional geology, summaries of work on oxygen isotope and molecular biological studies applied to the isthmus, and a series of chapters dealing with marine and terrestrial animals and plants whose evolution was affected by the formation of the isthmus.

Chapter Two. The Ocean Divided

There is unfortunately a dearth of popular literature concerning the natural history of the oceans bordering Central America. The nontechnical references given below address only a few of the topics discussed in chapter 2. The other references are from the technical scientific literature.

Nontechnical References

Allen, G., and R. Robertson. 1994. *Fishes of the Tropical Eastern Pacific*. University of Hawaii Press, Honolulu, Hawaii.

Carr, A. 1984. *The sea turtle: so excellent a fishe*. University of Texas Press, Austin.
A combination of superb natural history and delightful literary style. The best source of information on turtles and manatees and on the cultures of Caribbean peoples who made a living from the sea.

Carr, A. 1979. *The Windward Road*. University of Florida Press, Gainesville.
An account of southern Central American landscapes in the 1960s while the author was searching for turtle nesting sites.

Humann, P. 1994. *Reef Fish Identification: Florida Caribbean Bahamas*. Deloach, N. New World Publications, Jacksonville, Fla.

———. 1993. *Reef Coral Identification: Florida Caribbean Bahamas*. Deloach, N. New World Publications, Jacksonville, Fla.

———. 1992. *Reef Creature Identification: Florida Caribbean Bahamas*. Deloach, N. New World Publications, Jacksonville, Fla.
These are the most reliable and up-to-date popular guides for the identification of marine animals, particularly reef-associated species.

Technical References

Jackson, J. B. C. 1991. *Adaptation and Diversity of Reef Corals*. Bioscience 41 (7): 475–82.
This article was written for biologists in all fields and is thus somewhat less technical than one in a specialized journal. It discusses how coral reefs are

similar to tropical rain forests in their diversity of species and community structure. The particular importance of the role of such catastrophes as hurricanes and epidemic diseases is also described.

Jackson, J. B. C., A. Budd, and A. G. Coates (Editors). 1996. *Evolution and Environment in Tropical America.* University of Chicago Press, Chicago.
Several chapters present recent research on the ecological and evolutionary responses of marine animals, over the last 10 million years, to the rise and closure of the Central American isthmus.

Jones, Douglas S., and P. F. Hasson. 1985. "History and Development of the Marine Invertebrate Faunas Separated by the Central American Isthmus." In Stehli, F., and S. D. Webb (Editors), *The Great American Biotic Interchange,* 325–55. Plenum Press, New York.
A good review of the marine paleontological evidence, as of the early 1980s, for the rise and closure of the Central American isthmus.

Keller, B. D., and J. B. C. Jackson (Editors). 1993. "Long-term assessment of the oil spill at Bahía Las Minas, Panamá," Synthesis report, volume I: Executive summary. OCS Study MMS 93–0047. U.S. Department of the Interior, Minerals Management Service, Gulf of Mexico OCS Region, New Orleans, La. 129 pp.
Comprehensive scientific review of the ecological effects of a major oil spill on coral reefs and mangroves of the Caribbean coast of Panama based on six years of research following the spill and more than ten years of monitoring the reef before the spill. (There is an excellent executive summary in both Spanish and English.)

Chapter Three. Central American Landscapes

Beebe, W. 1942. *The Book of Bays.* Harcourt Brace, New York.
An account of a voyage down the Pacific coast of Central America in the 1930s. A good description of the coastal landscapes, sometimes with excursions inland as well as comments about marine life.

Belt, T. 1985 (Reprint). *The Naturalist in Nicaragua.* University of Chicago Press, Chicago.
A classic pioneering study of tropical natural history written in the 1870s. Contains evocative descriptions of typical Central American Landscapes. Required reading for any naturalist in Central America.

Benzoni, G. 1857. *The History of the New World.* The Hakluyt Society, London.
A lively, perceptive account of Central American travels by an Italian adventurer who participated in the conquest.

Boza, M., and R. Mendoza. 1981. *The National Parks of Costa Rica.* INCAFO, Madrid.
A thorough and profusely illustrated description of the National Parks of Costa Rica. Contains much information that is relevant to other countries.

Carr, A. 1979. *High Jungles and Low.* University of Florida Press, Gainesville.
Vivid and ecologically informed descriptions of northern Central American landscapes ranging from the mountains to the lowland forests. This book was written in the 1950s, so some of its natural history information is now dated.

D'Arcy, W., and M. Correa. 1985. *The Botany and Natural History of Panama.* Missouri Botanical Gardens, St. Louis.
A compendium of articles on Panamanian ecology and ethnology.

Janzen, D. (Editor). 1983. *Costa Rican Natural History.* University of Chicago, Chicago.
A detailed compendium of species and ecosystems in Costa Rica, applicable in many ways to other Central American countries also. Although somewhat deficient in aquatic habitats and species, it provides articles on geology and paleontology.

Kricher, J. 1989. *A Neotropical Companion.* Princeton University Press, Princeton.
A general overview of tropical American ecology that has frequent references to Central America.

Moser, D. 1975. *Central American Jungles.* Time-Life, New York.
A clearly written and well-illustrated survey of Central American natural landscapes and wildlife, ranging from the Guatemalan Highlands to the Darién. Although a little dated, it is still an excellent introduction to the region.

Sanderson, I. 1941. *Living Treasure.* Viking Press, New York.
A lively account of a biological expedition to Belize in the late 1930s.

Stephens, J. L. 1969 (Reprint). *Incidents of Travel in Central Chiapas and Yucatán.* Dover Publications, New York.
A fascinating insight into Central American society and politics during the early 1800s from a detailed account of an expedition that pioneered Maya archeology. This classic saga contains many vivid, although sometimes biologically ill-informed, descriptions of regional landscapes.

Wallace, D. R. 1994. *Adventuring in Central America.* Sierra Club Books, San Francisco.
A travel guide, concentrating on Central American landscapes; includes basic natural historical descriptions, travel information, and maps.

Wallace, D. R. 1990. *The Quetzel and the Macaw: The Story of Costa Rica's National Parks.* Sierra Club Books, San Francisco.
An informal history that stresses how Costa Rican conservationists tried to include examples of all the country's ecosystems in its parks.

Chapter Four. The Great American Faunal Interchange

Eisenberg, J. F. 1989. *Mammals of the Neotropics: The Northern Neotropics.* Volume 1. University of Chicago Press, Chicago.
An authoritative and well-illustrated treatment of the living mammals of Central and northern South America.

Laurito, C. A., W. Valerio, and E. Vega. 1993. "Nuevos Hallazgos Paleovertebradológicos en la Península de Nicoya: Implicancias Paleoambientales y Culturales de la Fauna de Nacaome." *Revista Geológica América Central* 16:113–15.
This paper provides valuable data on Pleistocene vertebrates and interpretations of their environment on the Pacific slope of Costa Rica.

Simpson, G. G. 1980. *Splendid Isolation: The Curious History of South American Mammals.* Yale University Press, New Haven.
This readable review is the last word on the subject by the world's authority. It includes a discussion of the interchange of land faunas between North and South America.

Stehli, F., and S. Webb. 1985. *The Great American Biotic Interchange.* Plenum Press, New York.
This multiauthored work brings together most of the geological and biological evidence about the exchange of land faunas between North and South America after the closure of the Isthmus of Panamá.

Webb, S. D. 1991. "Ecogeography and the Great American Interchange." *Paleobiology* 17 (3): 266–80.
This review considers the ecological patterns of the extinct and living vertebrates that were involved in the Great American Interchange.

Webb, S., and S. Perrigo. 1984. "Late Cenozoic Vertebrates from Honduras and El Salvador." *Journal of Vertebrate Paleontology* 4:237–54.
This article provides detailed documentation of many of the richest Central American fossil vertebrate sites relevant to the Great American Interchange.

Chapter Five. The History of Forests on the Isthmus

Bush, M., D. Piperno, P. Colinvaux, L. Krissek, P. de Oliveira, and M. Miller. 1992. "A 14,300-year paleoecological profile of a lowland tropical lake in Panama." *Ecology Monographs* 62:251–75.
A detailed and comprehensive technical description of the climatic and vegetational history of Lake La Yeguada.

Bush, M., and P. Colinvaux. 1990. "A long record of climatic and vegetation change in lowland Panamá." *Vegetation Science* 1:105–19.
The technical paper describing the research on the long core from El Valle that is described in this chapter. This core represents the longest continuous section of Ice Age sediments yet obtained from the tropical lowlands of the New World.

Colinvaux, P. A. 1993. "Pleistocene Biogeography and Diversity." In *Tropical Forests of South America,* P. Goldblatt (Editor). Yale University Press, New Haven.
A review of all the data for reconstructing climate and vegetation in the New World Tropics, including those for Panama presented here but with special focus on the Amazon.

Colinvaux, P. A. 1989. "The Past and Future Amazon." *Scientific American* 260: 101–08.
A major review of the long debate as to the climate and type of vegetation of the Amazon (and by implication Central America) during the Ice Age. The arguments for the Amazon lowlands being arid, with savanna vegetation, are summarized. An alternative view is then put forward, arguing that the region remained forested in both Glacial and Interglacial phases, although the air temperatures were lowered by 6 degrees. This article was also published in Spanish, French, and German.

Flenley, J. 1979. *A Geological History of Tropical Rainforest.* Butterworths, London. Although now somewhat dated in technical details, this is still the best general account of the development of tropical rain forests from an earth science viewpoint.

Chapter Six. The Native Peoples of Central America

Abel-Vidor, S., and others. 1981. *Between Continents/Between Seas: Precolumbian Art of Costa Rica.* Harry N. Abrams, New York.
Lavishly illustrated compendium, with useful chapters on regional archaeology, stone sculpture, jade, and gold.

Fash, W. 1993. *Scribes, Warriors, and Kings: The City of Copán and the Ancient Maya.* Thames and Hudson, London.
A modern account of this important Maya City that stresses ecology and economy. It is also a useful account of the history of Maya research.

Flannery, K. 1982. *Maya Subsistence: Studies in Memory of Dennis E. Puleston.* Academic Press, New York.
Compiled in memory of a pioneer in studies of the Maya intensive agriculture, this book remains a very useful reference on past and present-day Maya agriculture and hunting.

Hammond, N. 1982. *Ancient Maya Civilization.* Rutgers University Press, New Jersey.
A very readable thematic approach to Maya archaeology.

Kelley, D. 1976. *Deciphering Maya Script.* University of Texas Press, Austin.
An excellent account of Maya writing, astronomy, and politics. Although rather detailed for the lay reader, it is highly recommended for those particularly interested in these aspects of Maya culture.

Lange, F. (Editor). 1992. *Wealth and Hierarchy in the Intermediate Area.* Dumbarton Oaks, Washington, D.C.
A collection of essays that contains useful, up-to-date information on early ceramics and the southern part of Central America.

Lange, F., and D. Stone (Editors). 1984. *The Archeology of Lower Central America.* University of New Mexico Press, Albuquerque.
A general survey of the archaeology of El Salvador, Honduras, Costa Rica, and Panama. Although now somewhat dated, there are very useful sections on geography and relations with South America.

Linares, O., and A. Ranere (Editors). 1980. *Adaptive Radiations in Prehistoric Panama.* Peabody Museum Monographs, Cambridge, Massachusetts.
This thorough review was much ahead of its time in taking an unusually ecologically oriented approach to prehistory. The book deals especially with agricultural origins and dispersals.

Porter-Weaver, M. 1982 (2d ed.). *The Aztecs, Maya, and Their Predecessors.* Academic Press, New York.
Now somewhat dated but still a very useful compendium for the Precolumbian archaeology of the Maya lands. It also contains a good summary of the archaeological data for México.

Sheets, P. 1992. *The Ceren Site: A Prehistoric Village Buried by Volcanic Ash in Central America.* Harcourt Brace Jovanovich, New York.
A detailed account of the houses, storerooms, kitchens, and workshops buried at this fascinating site.

Stone, D. 1972. *Precolumbian Man Finds Central America.* Peabody Museum Press, Cambridge, Massachusetts.
Now rather out-of-date but simply written, well balanced with respect to time and geography and excellently illustrated.

Chapter Seven. Spanish Rule, Independence, and Modern Colonization

The Effect of Geography

Stephens, J. L. 1969. (Reprint). *Incidents of Travel in Central America, Chiapas and Yucatán.* Dover Publications, New York.
The author was a lawyer, businessman, writer, daring traveler, and the father of Maya archaeology. A fundamental book for anyone wanting to know Central America in depth during the first half of the nineteenth century.

Wells, W. 1857. *Explorations and Adventures in Honduras, 1857.* New York.
A smooth, easy to read narrative by a North American traveler, describing the geography and customs of Honduras.

West, R., and J. Augelli. 1976. *Middle America: Its Lands and Peoples.* Prentice-Hall, Englewood Cliffs.
Deals clearly with the physical, political, and cultural geography of Central America.

Central America under Spain

Baron Castro, R. 1978. *La Población de El Salvador.* UCA Editores, San Salvador.
An outstanding and well-documented work on economic and sociocultural forces involved in the formation of El Salvador.

MacLeod, J. M. 1973. *Spanish Central America: A Socioeconomic History, 1520–1720.* University of California Press, Berkeley,
Necessary reading for any introduction to the subject. This erudite and global approach includes an excellent analysis of the role of indigo and cacao during the colonial period.

Mariñas Otero, L. 1987. *Honduras.* Universidad Nacional Autónoma de Honduras. Editorial Universitaria, Colección Realidad Nacional No. 6, Tegucigalpa.
A very readable panoramic view of the formation of Honduran society.

Martínez Peláez, S. 1973. *La Patria del Criollo: Ensayo de Interpretación de la Realidad Colonial Guatemalteca.* Editorial Universitaria Centroamericana, San José, Costa Rica.
A powerful account of the critical political dilemma of Guatemala: the relationships between indigenous peoples and mestizos.

Wortman, M. 1982. *Government and Society in Central America, 1680–1840.* Columbia University Press, New York.
 Particularly useful to understand the political and administrative framework of the colonial period and the first few decades of the republican period.

Independence and the Republics

Asturias, M. 1988 (3d ed.). *El Señor Presidente.* EDUCA, San José, Costa Rica.
 The classic novel of a Guatemalan Nobel Prize winner on the social psychology of a dictatorship. Based on the events of the four presidential terms of Manuel Estrada Cabrera at the turn of the century.

Burguess, P. 1926. *Justo Rufino Barrios—A Biography.* Dorrance and Company, Philadelphia.
 An account of the liberal revolution of the second half of the nineteenth century as seen in the action and political thinking of one of its foremost exponents, Barrios, the "Guatemalan reformer."

Perez Brignoli, H. 1985. *Breve Historia de Centro-América.* Alianza Editorial, Madrid.
 An excellent introduction to the region for the general public.

Torres-Rivas, E. 1973. *Interpretación del Desarrollo Centro Americano.* Editorial Universitaria Centroamericana. San José, Costa Rica.

Contemporary Dilemmas

Bulmer-Thomas, V. 1988. *The Economy of Central America since 1920.* Cambridge University Press, England.
 Explores the socioeconomic and political development, in five Central American republics, from 1920 to the lost decade.

Leonard, J. 1987. *Natural Resources and Economic Development in Central America: A Regional Environmental Profile.* International Institute for Environment and Development. Washington, D.C.
 Updated state of the environment and a call to change development strategies that are based on increasing depletion of natural resources.

Woodward, L. R. 1976. *Central America: A Nation Divided.* Oxford University Press, Oxford.
 Probably the only updated general history textbook on the region.

Modern Colonization Fronts

Heckadon-Moreno, S. 1984. "Panamá's Expanding Cattle Front: The Santeño Campesinos and the Colonization of the Forests." Ph.D. diss., University of Essex, England.
 A detailed recent account of the development of the colonization fronts of Panama.

Schwartz, N. B. 1990. *Forest Society: A Social History of Petén, Guatemala.* University of Pennsylvania Press. Philadelphia.
 A comprehensive social and economic history of Guatemala's most dynamic colonization front.

Chapter Eight. Central American Indians Today

Adams, Richard N. 1956. "Cultural Components of Central America." *American Anthropologist* 58:881–907.

A fundamental essay on the population components and cultural traditions of contemporary peoples and cultures of the region, including the indigenous groups.

Bozzoli de Wille, María E. 1986. *El Indígena costarricense y su ambiente natural: Usos y adaptaciones.* San José, Editorial El Porvenir.

This small volume presents an overview of Costa Rica's indigenous groups and discusses their relationships with the natural environments found on their territories.

Chapin, M. 1992. "The View from the Shore: Central America's Indians Encounter the Quincentenary." *Grassroots Development* 16 (2): 2–10.

Chapin reviews the status of indigenous peoples in Central America, focusing on their relationship to the surviving regional biodiversity and their attempts to organize to defend their lands and rights.

Conzemius, E. 1932. *Ethnographical Survey of the Miskito and Sumu Indians of Honduras and Nicaragua.* Smithsonian Institution, Bureau of American Ethnology, Bulletin 106. United States Government Printing Office, Washington, D.C.

This is a classic ethnographic study and baseline source on Miskito and Sumu life in Honduras and Nicaragua during the early twentieth century.

Cruz, F. 1984. "Los indios de Honduras y la situación de sus recursos naturales." *América Indígena* 44 (3): 423–45.

The Honduran anthropologist Cruz describes the position of Honduras's Indian population and their lands in relation to state laws and regulations. He considers their cultural identity, legal land rights, and resource use in the face of conservation interests and other development forces that threaten their land security.

Davidson, W. 1987. "The Amerindians of Belize, An Overview." *América Indígena* 47 (1): 9–21.

The geographer Davidson's overview of the cultural history of the indigenous populations of Belize has discussions of the Mopán, Kekchí, and Yucatec Maya, focusing on their access to lands and position in the national society.

Davidson, W., and Melanie Counce. 1989. "Mapping the Distribution of Indians in Central America." *Cultural Survival Quarterly,* Chapin, M. (Editor), 13 (3): 37–40.

The authors provide an accurate assessment of the distribution of indigenous populations in the region for the early 1980s. Their map is the centerfold in this special edition of Cultural Survival Quarterly which includes articles on a variety of indigenous groups in Central America.

Dow, J., and R. Kemper. 1995. *Encyclopedia of World Cultures.* Volume 8. *Middle America and the Caribbean.* G. K. Hall/Macmillan, New York.

This volume provides up-to-date cultural overviews of the indigenous peoples of Central America written by scholars actively researching in the region.

González, Nancie. 1988. *Sojourners of the Caribbean: Ethnogenesis and Ethno-history of the Garífuna.* University of Illinois Press, Urbana.

This is the definitive work on the Black Carib, or Garífuna, populations of Central America, including their cultural history and identity.

Gordon, B. 1982. *A Panama Forest and Shore: Natural History and Amerindian Culture in Bocas del Toro.* Boxwood Press, Pacific Grove, California.

This is an important contribution on the cultural ecology of contemporary Indian subsistence. Gordon details how forests (primary and secondary) with their plant and animal life are influenced by the indigenous Guaymí of western Panama, describing the details of a time-tested subsistence system that depends on the maintenance of natural environments.

Helms, Mary. 1971. *Asang: Adaptations to Culture Contact in a Miskito Community.* University of Florida Press, Gainesville.

This landmark ethnographic study details village life of the Miskito people in Asang, a community along the Río Coco in Nicaragua. Helms uses ethnohistorical data combined with primary fieldwork to detail how the social, economic, political, and religious life of the Miskito have been influenced by contact with outside agents of change.

Herlihy, P. 1992. "Wildlands Conservation in Central America during the 1980s: A Geographical Perspective." In T. Martinson (Editor), *Geographic Research on Latin America: Benchmark 1990.* Yearbook, Conference of Latin Americanist Geographers, No. 17–18;1–43.

This article focuses on the issues revolving around the conservation of natural resources in protected areas inhabited or exploited by indigenous and other resident populations.

Howe, J. 1986. *The Kuna Gathering: Contemporary Village Politics in Panama.* University of Texas Press, Austin.

This ethnographic study details the Kuna political system in Panama. The volume helps us understand that community organization is at the core of how Kuna people have been able to defend their lands and culture while confronted with outside national and international pressures.

Incer, J. 1990. *Nicaragua: Viajes, Rutas, y Encuentros, 1502–1838.* Asociación Libro Libre, San José, Costa Rica.

This important volume examines the geographical discoveries of explorers, missionaries, colonists, and scientists to provide an assessment of the lands and cultures of Nicaragua, especially its indigenous peoples.

Lovell, G., and C. Lutz. 1995. *Geography and Empire: A Guide to the Population History of Spanish Central America, 1500–1821.* Westview Press, Boulder.

This volume provides a valuable survey of recent literature on the population history of colonial Central America. An introductory chapter looks at the native population size at contact time and discusses the demographic decline and changes in the colonial population due to additions of African slaves, Spaniards, and mixed groups. The volume then annotates more than two hundred sources dealing with population research on the region, much concerning the indigenous peoples.

Miller, M. 1993. *State of the Peoples: A Global Human Rights Report on Societies in Danger.* Beacon Press, Boston.

This volume supplies accurate, up-to-date information on the status of indigenous peoples around the world with an important section devoted to Central America. The volume includes a feature article on the Tawahka Sumu struggle to secure their homelands in the Honduran Mosquitea region.

Nietschmann, B. 1973. *Between Land and Water: The Subsistence Ecology of the Miskito Indians, Eastern Nicaragua.* New York, Seminar Press.
This field study in geography focuses on the ecology of Miskito subsistence as it relates to the ecosystems on the Atlantic coast of Nicaragua. Nietschmann uses an ethnoecological approach and focuses on the village of Tasbapauni at Pearl Lagoon but includes much useful background on the culture history and social organization of the Miskito.

Steward, J. H. (Editor). 1963. *The Circum-Caribbean Tribes.* Volume 4, *Handbook of South American Indians.* Smithsonian Institution, Bureau of American Ethnology Bulletin No. 143. Cooper Square Publishers, New York.
This volume is a major source on the indigenous peoples of Central America. A number of contributions from leading scholars, including Frederick Johnson, Doris Stone, Paul Kirchhoff, Samuel Lothrop, and David Stout, provide important overviews of the geography and history of specific native groups in the region.

Torres de Araúz, R. 1980. *Panamá Indígena.* Panamá, Instituto Nacional de Cultura, Patrimonio Histórico, Impresora de la Nación, Panama City.
This is a comprehensive ethnography of Panama's indigenous populations, including a review of the past research, ethnohistory, populations, and culture of each group.

Chapter Nine. The Paseo Pantera Agenda

Barzetti, V. (Editor). 1993. *Areas Protegidas y Desarrollo Económico en América Latina y El Caribe.* World Conservation Union (IUCN) and the Interamerican Developmnent Bank, Washington, D.C.
A comprehensive book that addresses the major problems faced in Latin America and the Caribbean for the management of protected areas. Particular emphasis is placed on the role that appropriate management of the protected areas can play in economic development.

Central American Alliance for Sustainable Development. 1994. Central American Commission on Environment and Development and the Inter-American Institute for Cooperation on Agriculture. San José, Costa Rica.
Available in both Spanish and English, this publication contains all the key documents, adopted at the summits of the Central American Presidents, that comprise the Central American Alliance for Sustainable Development.

Jukofsky, D., 1992. "Path of the Panther." *Wildlife Conservation* 95, no. 5 (September/October).
A good review of the development of the concept of the Path of the Panther (Paseo Pantera). It particularly stresses the need to improve management and development of key protected areas throughout Central America.

Munasinghe, M., and J. McNeely. 1994. *Protected Area Economics and Policy: Linking Conservation and Sustainable Development.* The World Bank and the World Conservation Union (IUCN), Washington, D.C.

An excellent review of how protected areas can contribute to sustainable development. A collection of articles by experts in the respective fields describes case studies involving economic incentives, economic valuation of protected areas, ecotourism, and funding mechanisms.

Stein, E., and A. P. Salvador, (Editors). 1992. *Democracia sin Pobreza: Alternativa de Desarrollo para el Istmo Centroamericano.* Editorial Departamento Ecuménico de Investigaciones (DEI), San José, Costa Rica.

A collection of articles on the political and economic challenges that face Central America in its quest to achieve lasting peace and socioeconomic development.

Westing, A. H. (Editor). 1993. *Transfrontier Reserves for Peace and Nature: A Contribution to Human Security.* United Nations Environment Programme (UNEP), Nairobi, Kenya.

Although this volume is directed primarily at Indochina, it contains a useful discussion of the general need and particular value of maintaining peaceful and harmonious political conditions, through joint conservation development, across national borders.

Index